The Longest Bench

The Longest Bench

Down and Out in Brighton

TK GONDO

First Edition

Dedication

To my family, who I didn't tell about this because I didn't want to worry them. I deeply encourage you not to read this, but if you do, thank you for your understanding in advance.

To Krissie and Ivy, thanks for keeping me (relatively) sane through it all.

To Rory, Alex, Callum, Dmitry, Lewis, Monte, Farah, Bibby, Dempsey, Jacob, Dan, Yoann, Ana, Lorna, Ryan, Frederica, Jim, The Pinkies, Leo, Joel, Sammy, Angelo, Morgan, Jacob, Tiberius Zach, and Zee. You know what you did.

To Mrs. Reynolds, Mr Standing, Alex Hourston and Mr. Jenkins, I write because of you.

To The Clock Tower Sanctuary and all they do to help the down-and-out of Brighton.

"It is a feeling of relief, almost of pleasure, at knowing yourself at last genuinely down and out. You have talked so often of going to the dogs — and well, here are the dogs, and you have reached them, and you can stand it. It takes off a lot of anxiety." — George Orwell

The Crucible

The long bench stretched behind Brighton Pavilion like some accidental line in the sand, a boundary marker that no one had consciously drawn. On one side, neatly trimmed lawns, Regency architecture, and overpriced lattes; on the other, the messy sprawl of human misfortune—loud, raw, and uncontained.

The bench itself was an unremarkable slab of wood and metal, battered by time and weather. But what it lacked in aesthetics, it made up for in stories. Drama, comedy, heartbreak—if you knew where to look, the bench had it all, a theater of the absurd playing nightly, with tickets free of charge.

The police, when filing their reports, referred to the area's inhabitants as "Brighton's street community." This was a bureaucratic euphemism, of course. I doubt they called them anything so polite once their body-worn cameras were off.

No, they likely used the same words they used for themselves—nitties, smackheads, crackheads, and riff-raff, though "useless" and "scroungers" also came up often in muttered insults from passers-by.

The bench was where I landed, a kind of homecoming to a place I'd never intended to visit. I ended up there through a predictable series of steps: poor financial decisions, too many "one more" nights in the pub, and a few doomed attempts at romance with well-meaning women who had far more patience than I deserved. The people on the bench were no better or worse than I was; they had simply gotten there first.

By the time I joined them, I had shed whatever lingering notion I had of being "above" their situation. The truth is, you adapt quickly to the bench—its rhythms, its rules, and its peculiar hierarchies. You have to. The bench doesn't care who you are or where you came from. It's not a haven. It's a crucible. It strips away the unnecessary, the indulgent, the soft edges of who you thought you were, leaving behind something harder, leaner, and infinitely more aware.

The bench itself was a watering hole, a bed, and a battleground. At night, lost souls would lie on its hard planks with their coats tucked under their head and their rucksacks clutched tightly to my chest. The mornings were a blur of bleary eyes, dry mouths, and the eternal question of what to do with the day.

There were unspoken laws, of course: don't take someone's spot, don't touch someone's bag, and don't ask too many questions unless you're prepared to hear the answer. The longer you stay, the more you realise the bench is not just a piece of furniture; it's a crucible. It tests you, wears you down, and reshapes you into something harder, leaner, and perhaps a bit less human.

The inhabitants of the bench were my companions and, in their own way, my teachers. To the outside world, we were invisible unless we made ourselves unavoidable. I learned quickly that standing out on the streets is a double-edged sword. I'm a big man, loud and noticeable, and this has never been an advantage out here. Attention invites trouble, and trouble on the bench is never far away.

And yet, for all its chaos, there was a strange dignity to life on the bench. Not the kind of dignity that earns applause or recognition, but the quiet dignity of survival. Of shared suffering. Of understanding, without needing to say it, that everyone here was carrying their own battles, their own weight. The bench had a way of revealing truths you didn't know you were hiding. It stripped you down, but sometimes, what it left behind was something you didn't realise you were capable of: endurance, resilience, and maybe, just maybe, a little hope.

Brighton in 2024

Brighton in 2024 is a city that wears its contradictions like a tailored jacket—a little too snug, with frayed seams hinting at the strain beneath. It's a place that markets itself as a haven for artists, progressives, and free spirits, where rainbow flags flutter proudly in the salty sea breeze and vegan cafes outnumber kebab shops.

The city's nickname, "London by the Sea," feels both apt and ironic. It boasts the cosmopolitan air of the capital, but with the undercurrent of something rawer, grittier, and undeniably precarious.

At first glance, Brighton is a postcard come to life. The Regency architecture stands as a monument to an era of indulgence, its terraces and crescents now housing everything from boutique hotels to overpriced Airbnbs.

The Brighton Pavilion, an ornate fever dream of domes and minarets, anchors the city's historical identity, though its grandeur feels somewhat out of place amid the city's modern struggles. The seafront hums with life—tourists lining up for fish and chips, locals jogging along the promenade, the rattle of the Volk's Electric Railway trundling along the pebbles. It's the Brighton that makes the brochures, the one visitors snap selfies with before hopping on the train back to Victoria.

But the Brighton of 2024 isn't just its polished façade. Beneath the bohemian veneer lies a city grappling with the same crisis plaguing the rest of the country, amplified by the peculiar pressures of being a seaside refuge. The housing crisis here feels particularly acute.

 Rents have spiraled into the stratosphere, forcing out the very creatives and misfits who once gave Brighton its charm. The city's population has swelled, not just with Londoners seeking a slower pace, but with those fleeing the grinding monotony of small-town life, hoping to reinvent themselves by the sea.

Brighton, like much of Britain, is living through the long hangover of Brexit, the COVID-19 pandemic, and a government that seems hell-bent on austerity as a philosophy.

The cost-of-living crisis looms large. Energy prices have turned winter into a season of dread, even in a city as liberal as Brighton, where communal spirit and resourcefulness are often touted as antidotes to systemic failure. Food banks are no longer a stopgap—they're a cornerstone of survival. And the streets? They've become a stage for the human fallout of all of it.

Homelessness here isn't hidden away. It's visible, omnipresent, woven into the fabric of the city. Brighton's benches, doorways, and sheltered corners are claimed nightly by those with nowhere else to go. The council struggles to keep up, its resources stretched thin by rising demand and dwindling budgets.

Emergency hostels, often no more than holding pens for despair, are overflowing. People live in tents along the edges of parks, forming makeshift communities that the authorities periodically dismantle, as though out of sight could mean out of mind.

Yet, Brighton's contradictions shine through even in its darkest corners. The same city that struggles to house its most vulnerable also hosts one of the country's largest Pride celebrations, a kaleidoscope of joy, defiance, and resilience. The same streets that witness addiction and violence also pulse with creativity—buskers, graffiti artists, and writers documenting it all in real-time. Brighton remains a city of expression, even as its people fight for space to exist.

The city's relationship with its homeless population is fraught, to say the least. Brighton brands itself as inclusive, yet the systemic neglect of its most marginalised residents tells a different story. Community Protection Notices (CPNs) are handed out like confetti, criminalising poverty under the guise of maintaining public order. The police sweep the streets of visible homelessness before tourist-heavy weekends, a desperate attempt to preserve the illusion of Brighton as a carefree, seaside utopia.

The local economy, heavily reliant on tourism, exacerbates the tension. Brighton thrives on the promise of escapism, of weekends spent drinking overpriced cocktails with a view of the sea or rummaging through vintage shops in the Lanes. For those of us on the bench, watching the city's transitory population—people who come here to play but never stay—is like standing outside a glass window, looking in on a party we'll never be invited to.

And then there's the pier, Brighton's crown jewel, stretching out into the Channel like a beacon of nostalgia. Its bright lights and arcade sounds attract tourists like moths to a flame, but for those of us living on the streets, it's a symbol of everything we don't have. The pier is where people go to feel weightless, to forget their problems for a while. We can't forget ours. We see them reflected in the oily black waters below, endless and unyielding.

But Brighton is still Brighton, a place where rebellion bubbles beneath the surface. Activists take to the streets with banners demanding housing reform, fair wages, and climate action. The Green Party holds a shaky dominance over the city council, a rare flicker of progressive politics in a country otherwise gripped by conservatism. Small victories—a new homeless shelter here, a public art installation there—remind you that people are still trying, still fighting, still believing in something better.

Brighton in 2024 is a city on the edge, a microcosm of the country's struggles but also its stubborn hope. It's a place where despair and beauty coexist, where the fight for survival is met with moments of unexpected grace. It's where I ended up, where the bench became my world, and where I learned that even in the most fractured places, life finds a way to persist.

"What sort of chronic saddo really believes that the best days of his life were spent in the mud at Woodstock or fighting on Brighton beach?" — JULIE BURCHILL

Fighting in the Streets

It will come as no surprise to you that the streets are as violent as they are cold, as they are damp and lonely and motherless and unforgiving. Blood flows freely here. Falling face-first into the gravel at the hands of some toothless lunatic with a grudge is as common as rain in Glasgow.

Once upon an immature version of me, I used to brag that I'd never been punched in the face despite the fact that I have the ideal personality for face-punching. Then came the psycho who had recently just come out of prison who thought I was gay (I'm not) because I'm not good at personal space. He gifted me with two well-placed punches above my left eye that felt like a kiss from a northbound train and gave me what I'd been bragging about not receiving for so long. This was the first time I received *the good news*.

Street fights are the darkly comic cousin of bad reality TV. They have all the intensity, all the drama, but none of the cameras, none of the cutscenes, none of the edits to make it watchable. The most common one was suspicion of theft. Homeless Person X would swear on his last penny that Homeless Person Y has stolen their *spice* —the only thing dulling the edges of his sharp reality.

X kicks up an unbelievable fuss and sprays Y with profanity and blood-curdling threats if they don't hand over the alleged stolen synthetic intoxicants. Y refuses to back down and is almost certainly bluffing like a bad poker player. He vehemently denies the accusations but it's already too late because X has the look of the devil in his eyes and wants to inflict pain on Y more than they want their shit drugs back.

The situation is now already out of control. In my first few days on the streets, I would jump in and use my received pronunciation accent to be the voice of reason- to rationalise with the irrational. After taking more than a baker's dozen of stray punches in the crossfire I've learned that this is a fool's errand and I am only a fool when it suits me best. Now with experience of this cruel way of life under my belt, I sit back and watch with scrump cider and no popcorn. Punches are thrown, integrities are insulted and it's over as soon as it begins.

The aggressor generally tends to win. Here people only pick fights they *think* they can win. Y storms off in a huff and puff, humiliated by the unnecessary affair. About 30 minutes later, X realises that they have more pockets on their multiple layers of winterproof clothing than they thought they had. The bag of drugs is found, buried deep in some pocket he forgot existed.

Y never finds out and I return to the bench the next day to see them laughing and joking with each other as if yesterday was something I imagined. This is the street—conflicts rise and fall like bad waves on a polluted shore, and if you're lucky, you live to float another day.

X and Y have both been birthed through the bloody canal of the British care and foster system. This has not equipped them with the emotional intelligence to deal with their petty disputes with any means other than unarmed combat. When they turned 18, the state closed its doors on them like an angry landlord, tossing them to the wolves. Cue the endless cycle: drifting through days that blur into weeks, which smear into years, until trauma rears its ugly head and finally gives them a reason to get noticed—usually in the form of a crime. That's when they get their ticket to the finishing school of His Majesty's Prisons, where the lessons are few but the scars are plenty.

The code of chivalry is rarely adhered to on the streets. Women hit men and men hit them back. Regardless of how many times I shouted "You can't hit a woman mate" it tended to do sweet fuck all to prevent the violence. Instead, people would clap back at me and say something along the lines of "If they want equal rights, they should get equal lefts"- this is ironically a phrase I heard blurted out by women more than anyone else.

Men often used the threat of violence more than they inflicted it on their counterparts of the fairer sex. The bench taught me that if you get really good at making creative threats against people's safety and health you never have to use your fists. As the saying goes, power only works until it is used.

Violence here isn't just physical; it's psychological, too. Threats and intimidation are an art form, and some of the people on the bench are Michelangelo with their words. "I'll carve you up like a Christmas turkey," one bloke shouted, holding a butter knife he'd nicked from Wetherspoons.

The ridiculousness of it would be funny if it wasn't so real. The knife doesn't have to be sharp, and the threat doesn't even have to be plausible—what matters is the performance. If you can make someone believe you're unhinged enough to follow through, you've won.

The rules of engagement are flexible, to say the least. Weapons are a last resort, but I've seen everything from bike chains to broken bottles get used when things spiral out of control. People don't carry knives out of malice—they carry them out of fear. Fear of the next fight, fear of being the one who doesn't get up after the first punch. There's no referee on the streets. If you go down, you stay down until someone decides they're done with you.

And then there are the spectators—people like me, who've learned that stepping in is a fool's errand. We sit there, silent or sipping on something strong, watching the drama play out like a grim reality show. No one cheers. No one eggs it on. The fights are just another part of life on the bench, like the rain or the police making their rounds. We've all taken our hits, and we've all given a few back. It's a shared understanding, unspoken but deeply felt: what happens in the fight stays in the fight.

But not everyone gets to walk away. I've seen punches connect in ways that make your stomach churn, seen heads crack against pavement with a sickening thud. The thing about the streets is that the line between a scuffle and a tragedy is razor-thin, and it's always lurking, waiting for the wrong punch at the wrong time. One bad fall, one unlucky hit, and that's it. No hospital, no ambulance—just someone shouting, "Oi, you alright?" and praying for a grunt in response.

For all its chaos, the fighting reveals a harsh truth about the streets: this is a place of survival, not living. The violence is just a symptom, a side effect of lives spent at the edge of the abyss. On the surface, it's about pride or stolen spice, but underneath, it's about desperation. It's about people who've been told, over and over again, that they're nothing. Out here, every punch is a way of saying, *I'm still here. I still matter.*

And maybe that's why, for all its ugliness, you can't fully condemn it. The fights are brutal, yes, but they're also the last shred of control many of these people have. When the world has taken everything else, the ability to throw a punch—or take one—is a reminder that you're still alive. It's messy, it's tragic, and it's far from noble, but it's real. And on the bench, real is all you've got.

The most talented purveyor of violence I met on the streets was an Irish traveller. I dare not say his name because I don't want to wake up dead in the moors tomorrow but let us call him Paddy for argument's sake. If his circumstances were different, Paddy would be Conor Mcgregor.

He had the aura, the charisma, the machismo, the fearlessness and the fists, knees, elbows and tattooed forehead to back it up. I trace the beginning of my battle with PTSD to one particular night. He stood by the bench, a group of Frenchmen were getting overconfident because they had the advantage in numbers.

What happened next will live in my memory until my last day. Paddy knocked them out practically one by one. One of them woke up from his head kick-induced slumber and looked to catch him from behind. Paddy noticed him in his peripheral vision and fucking overhead bicycle kicked him. A fucking bicycle kick.

He is the first person I have ever met that I can genuinely say feared no man. He didn't give a shit if someone was a foot taller, or broader- he would always stand and fight. In the same way that people ask joggers if they are running away from something or towards something I would ask myself (and never him for obvious reasons) if he was actually fighting the person in front of him or demons inside himself.

When I got to know his story it was obvious that the answer was the latter. Although he was still a very young man, he, like many other members of the bench, had spent his formative years in prison for accidentally killing another traveller teenager with a punch in a streetfight. He was reared into manhood by the prison guards and hardened criminals.

There was a sense of inevitability to him becoming a warrior on the streets. As a boy, before he would be thrown into a fight by the palms of his father, he would whisper "You better not disgrace my family name or you won't be coming back home tonight". How could he not become the person he became?

"When I get tired of London, I go to Brighton. When I get tired of Brighton, I go to bed."—Clemence Dane

Liam

I can't remember the first time I met Liam, but that's irrelevant—once you've met someone like him, they're burned into your memory, permanent as a bad tattoo. He was a good-looking bastard. He had these piercing bright electric blue eyes that were made sharper by the contrast of the black guyliner he smeared on like warpaint.

He was blessed with a full head of shaggy rockstar hair which was long enough to make him look like the fifth, missing member of *The 1975* but too short to be described as long. His energy was endless, and relentless, like he was on a mission to burn through life faster than the rest of us. He was magnetic in a way that made people forgive the obvious fact: Liam was annoying as sin.

No, not just *annoying*—driving-you-up-the-wall insane. He'd get under your skin, press every nerve until you'd had enough, and just when you thought you'd ditch him, he'd do something that made you forget all his past crimes until the next one. He had an energy that made him hard to like, hard to hate, and absolutely impossible to ignore.

Early one morning, I jogged down to the beach to play basketball before the courts got busy. It was one of those days you could tell was going to be gorgeous, and I wanted Brighton all to myself before everyone else woke up and caught on. The water was filthy, cold enough to shock some life into me, but I swam anyway.

I took a cold shower on the beach, rehydrated, and made my way uptown. As I strutted down the New Road with mid-summer day happiness, I heard a voice shriek my name to the clouds. *What the fuck have I done now?* I thought. I turned around to see Liam barreling toward me, looking like he'd just broken out of an asylum with enough energy to power the home counties. I was pleased to see him.

"*Come with me to the pharmacy*," he said, not really asking, just pulling me along, and I thought, *why not?*

We talked about sweet fuck all on the short walk, but whatever it was we discussed, it was hilarious because I was in stitches. It was easy with him; there was this natural absurdity to our friendship. Then we got to the pharmacy, and reality snapped back like a rubber band.

A lineup of obvious addicts was already outside, bodies worn down by whatever poison had gripped them, all of them clinging to the building like barnacles. I hadn't thought to question what Liam was picking up. After all, it was none of my business. *Hepatitis meds? Anti-anxiety pills? Athlete's foot?* I didn't care. But something didn't sit right.

"Liam, this isn't a *methadone* dispensary, is it?" I asked, the realisation dawning in my voice. Without missing a beat, he looked at me, dead serious, and said, "*Obviously*, mate. I said I had to get my prescription. I'm three weeks clean off *dark* now."

At this point, I'd learned not to give in to astonishment, and without hesitating, I put my hand on his shoulder and told him I was proud of him. I watched him down a sickly green cup of liquid hell, his fix in a measured, government-approved shot.

"If you took this, it would kill you, TK," he said with a smirk.

I shrugged. "Yeah, I'm good, thanks, mate."

As is my nature, I wanted to know how someone like him, young and with a life before him, wound up on a first-name basis with heroin. He told me, flat as an observation, that at 15, he was groomed and assaulted by some older boys who pretended to be his mates. They injected him with heroin and forced him into a life he'd been trying to escape ever since. He rattled off this horror like he was telling me what he had for breakfast.

But then he saw my face, saw that the grief in my eyes matched his, and his own façade cracked. We were just two young men, helpless, sitting on the edge of a park, crying over a broken world we couldn't change. I couldn't plaster over his cracks with words, and he couldn't erase my horror with his resilience.

Being around Liam was like playing with fire—you'd be laughing one second, in stitches over some ridiculous story, then in the next, he'd be shouting at strangers, his voice cutting through the air like a siren. It was like watching a storm roll in, sudden and chaotic.

At first, I didn't understand it; later, I'd know it for what it was—*withdrawal*. He'd snap at passersby, hurling insults at the top of his lungs, and that was my cue to get out. People might ask me why I even bothered with him. But who was I to judge? If anything, I think Liam liked having some piece of stability in his life in the shape of me, as much as that could be said for me, anyway. I wasn't exactly normal, but I wasn't *him*, either.

Then came the news. One day, someone sidled up to me, casual as you like, and asked, "TK, you're mates with Liam, right?"

I was expecting to hear some story about how Liam had gotten blackout drunk and annoyed someone and laughingly said, *"What's that cheeky cunt done now?"* What I heard next took me aback; they said he'd got into a fight and he was now on a life support machine.

I felt that horrible, icy grip of panic, but I tried to shrug it off as a rumour, a piece of the wild gossip that floats around Brighton like the seagulls. I'd soon be walking down the New Road in a few days to hear him shout my name. After all, this had happened once before when someone said that he'd been stabbed, only for me to see him walking down the street two days later.

It became very bloody clear that this was *not* the case when multiple people confirmed the rumour. It turned out to be worse than I thought. He was dead. Murdered in a petty squabble over God knows what, a senseless moment that took him away as if he was nothing. I screamed, wailed, and swung my bare fist into the nearest available wall before I remembered I was *not* hard.

The worst part of it all was my last interaction with Liam. He was being his usual annoying self and trying to rough play with people who just wanted to chill. My last words to him were, *"Liam, chill the fuck out or you're going to piss off the wrong person."* I hadn't meant it as a prophecy, but it felt like one.

Liam's death hit me like a sucker punch to the gut, not because it was unexpected but because it was inevitable in a way that I couldn't reconcile. I knew the world we lived in was cruel and careless, but it still felt surreal that someone as alive as Liam—someone so relentless, so chaotic—could be snuffed out over something meaningless.

His absence left a void, not just on the bench but in the warped ecosystem of Brighton's streets. People like Liam weren't meant to disappear quietly; they burned bright, burned fast, and left a scorched patch behind them.

What struck me most was how quickly his death became part of the background noise. On the bench, tragedy isn't an anomaly—it's a recurring event, like the changing tides or the seagulls that shriek at dawn. You mourn, you rage, but then you move on, because out here, grief is a luxury you can't afford for long. Survival doesn't wait for you to process your emotions.

Liam's story wasn't unique, and that's the most damning thing about it. For every Liam, there's a hundred more—men and women who were chewed up by systems that never cared for them in the first place. You start to see the patterns if you stick around long enough. They're predictable in their cruelty: abuse, addiction, incarceration, repeat. It's a cycle that feeds itself, trapping people like Liam in a loop that's nearly impossible to escape.

The streets are full of people who were failed long before they ever found themselves sleeping rough. The foster care system, the schools, the prisons—they're all part of the same machine, grinding people down and spitting them out onto the pavement. Liam's life was written in that cycle before he ever had a chance to choose otherwise. Groomed, abused, and abandoned, he didn't stand a chance.

And the worst part? No one seemed to care. Not really. Society only noticed Liam when he became a problem—when he was shouting in the street or picking fights in withdrawal. In death, he became another statistic, a fleeting headline, a ghost that would haunt no one but the people who shared his bench.

His death also brought into sharp focus the way the world looks at people like him. To most, Liam was invisible. A nuisance at best, a danger at worst. But to me, he was a reminder of the humanity that exists even in the darkest corners of life.

He was flawed, yes—deeply so—but he was also funny, loyal in his own unpredictable way, and desperate to claw his way out of the hole life had thrown him into. That desperation, that fire, is something the world doesn't acknowledge enough. It's easier to write people like Liam off than to confront the systems and circumstances that created them.

I thought a lot about the fight that killed him. I never found out the details—maybe it was over drugs, maybe over pride, maybe something even more trivial. Whatever it was, it wasn't worth it. But that's the thing about the streets: life becomes distilled down to its basest elements—survival, status, territory.

Violence is both currency and consequence, a language spoken fluently by those with no other recourse. Liam knew that language better than most. It was what got him through most days, and it was what took him in the end.

After Liam was gone, I noticed the cracks in the bench community more acutely. He was a disruptive force, yes, but he also had a way of holding things together, of breaking up the monotony with his antics. Without him, the air felt heavier, the nights colder.

I thought about how quickly he had faded from the minds of those who walked past us every day, and it made me angry. Angry at the indifference, at the systems that chew people up and spit them out, at myself for not doing more, for not being able to.

But anger doesn't fix anything, not out here. All it does is eat away at you, and I wasn't about to let it. Instead, I tried to hold on to the moments that made Liam who he was. His wild energy, his ridiculous sense of humor, the way he made me laugh even when I didn't want to. Those things didn't save him, but they mattered.

It's easy to rage against the system when you've seen it fail as spectacularly as it did with Liam. It's harder to figure out what could actually make things better. But sitting on the bench, night after night, surrounded by people trying to survive a world that seems determined to forget them, you can't help but imagine what might have been different.

For starters, the way we approach addiction is fundamentally broken. Liam wasn't a villain or a monster; he was a victim of trauma that he never had the tools to process. The system's answer to his suffering was to criminalise him, to shuttle him through prisons and probation offices without ever addressing the root cause. Addiction is treated as a moral failing rather than the illness it is, and that failure to see the human being behind the substance kills more people than the drugs themselves.

What if, instead of methadone clinics that feel more like cattle lines, there were safe spaces where people could detox and recover without judgment? Places staffed by people who understand addiction—not just academically, but personally—who know what it's like to stare into the abyss and claw your way back. What if we didn't wait until someone was at rock bottom to offer them help? Early intervention, trauma counseling, and real, ongoing support could save lives.

And what about housing? The idea that someone like Liam had to choose between a bed in a rat-infested hostel or the street is absurd. Housing first policies—giving people stable, secure homes before expecting them to tackle their other issues—have been proven to work.

When you have a roof over your head, suddenly everything else becomes just a little bit more manageable. You're not worried about where you'll sleep or if your things will be stolen, so you have the mental space to focus on getting better. It's not rocket science; it's basic human decency.

But even with housing, there's the question of what happens next. A home is a starting point, not a solution in itself. People need purpose. They need to feel like they're part of something, like they matter. Job training, education, mentorship programs—things that give people the tools to rebuild their lives—shouldn't be afterthoughts. They should be baked into the system, not handed out like rare prizes to the few who manage to jump through all the hoops.

And then there's community. One of the hardest things about being on the bench is how isolating it is. You're surrounded by people, sure, but there's a sense that you're all stranded together, cut off from the rest of the world.

That isolation breeds despair, and despair breeds more of the same cycles that keep people stuck. What if there were places where people could come together, not just to survive but to connect? Spaces where you're not judged for where you've been, where you can be seen as a person, not a problem.

Ultimately, the solutions aren't complicated. They're not flashy or revolutionary. They're about treating people like people—about recognising that everyone has a story, a history, a set of scars that brought them to where they are. It's about refusing to accept that some lives are worth less than others. The real challenge isn't figuring out what to do; it's finding the collective will to actually do it.

Because if there's one thing I know, it's that Liam's story doesn't have to keep repeating. It doesn't have to be inevitable. The systems that failed him can be changed; they just have to be seen for what they are first. And that means looking at people like Liam—not through the lens of pity or fear, but with the understanding that they are us, and we are them. Only then can we start to fix what's broken. Only then can we build a world where people like Liam don't have to burn out to be remembered.

I take solace in the fact that he told me he'd slept with three different women the night before he died. *Rest in peace, you annoying bastard.*

It's Raining
Cats and Geordies

Real love stories are rare in most walks of life. But the bench—this place of last resorts, where lost souls congregate—has seen its share. Love on the streets is everything at once: fiery, beautiful, violent, romantic, hateful, loving, overwhelming, requited, chaotic, beautiful, and romantic and loving and not what you thought it was going to be.

There is only one love on the long bench that I am in love with. The union of a woman named Cat and a man named Geordie. Who now love, and know enough to glimpse at the world through their thousand-yard stares. I hope they agree, but I understand why they clung to each other on that long bench.

Every cocky real and fake wannabe gangster loves to claim that they were *raised by the streets*. Cat can say this and say it with her chest. She became homeless for the first time when she was kicked out at the age of 13. Now she is at the age of 28 and has spent most of her life bedding down under stars, making nests in doorways, mattresses on benches, and cottages under bridges.

Although she made her best efforts to try and live a normal life complete with slow cookers, saturday night television, butter dishes, garden patios and pleasant elderly next door neighbours called Dave and Helen; she was part of the street and the street was part of her. Cat could comfortably sleep anywhere, a doorway, a tent, a bench- anywhere except a home.

This isn't for lack of trying. The hostels that the council would put her in were no place for a woman and no place for a man or dog. People would come and go to these graveyards for human dreams and get evicted for some misdemeanours or other. However, it was the rats, roaches, lice and bedbugs that were the only permanent tenants. She made the bold choice to be outdoors or in a flimsy tent, or rave squat with a 3-month deadline before police eviction; rather than suffer through insects and rodent flatmates.

The roach-hostel's loss appeared to be the streets of Brighton's gain. The streets gained a new matriarch. Cat—young as she was—was very naturally the Mother of the Bench. People older than her looked up to her. People who were motherless, orphaned and abandoned enough for that not to matter. She gives decent haircuts to people barbers would not look twice at, uses her first aid experience to bandage peoples booboos and shares the little that she has with those who need it more than she does.

Cat is, in no doubt, the queen of the long bench, a moniker most people would not want but one that she takes in her stride and does a remarkably good job. There's not a nook, cranny or character that is unbeknownst to her. There's a long grate that runs parallel to the front of the bench. One sunny, weirdly normal day, someone dropped their joint down into the grate. Inside the heart of darkness.

Personally, even if it was the last joint I would ever be allowed to smoke, I would let the grate keep the joint. This is on account of having seen half of Brighton, spit, vomit and worse into the grate. It is the heart of darkness. I went over to Cat who was sitting with her clever, good boy of a dog whose name is Boss. I told her what happened and she sprung into action. "Get me a couple of sticks TK" she said as she rearranged her hair.

I refused to believe she could do it, "Cat, you're surely not going to fish it out of there are you? I doubt a keyhole surgeon could do that". She looked at me dead in the eye and said "I've pulled fucking teeth from that grate". Moments later the disgusting joint was being smoked by its owner and I still haven't managed to scoop my jaw from the floor since.

Then there's Geordie. His name is not actually Geordie of course, but he's from Newcastle and people from Newcastle are known as Geordies, so everyone cleverly called him Geordie. His story was unique to me but so very typical of a man from the North East.

He grew up on an estate and spent the majority of his childhood in children's homes or being raised by carers. Being from Newcastle he was football mad and talented with his feet. He was signed to Hartlepool United as a young teenager and showed great promise until alcoholism sunk its teeth into his neck.

A cocktail of a traumatic childhood, discontent for the world he did not choose and a lack of great role models made developing the sickness inevitable for him. It wasn't long before he found it impossible to play without being drunk. The drink wrung the last bits of his talent dry.

He managed to turn it around by completing his secondary education and getting a business management degree whilst honing his skills as a consumer of Newcastle brown ale and chekov vodka. Things seemed to be going relatively well for him, for a boy who'd been through what he'd been through and seen what he had seen.

He looked like a modern viking and wrangled this into getting a modelling contract and not even for a local ale brewery- for actual clothing. The upward trajectory continued when he weaponised his geordie charm, business acumen and physical capabilities to start his own block-paving company.

He never told me what it was that brought him to the long bench of Brighton, but I can only assume it had something to do with the drink and the council getting in between him and his children. After a number of years being on the streets of London doing only God knows what he found his way to Brighton and his place on the longest bench.

On the long bench all days blur into one. Yesterday could have been a week, a fortnight or a month ago- but it definitely happened. This is why I have a hard time remembering how Cat and Geordie first got together. One day they were strangers and in less than a few days they were an item.

The streets accelerate all relationships- whether they be platonic or romantic. In one day of knowing someone you can experience 10 times the trauma, adventure and danger you would share with *friends* in the *normal* world. As someone with a penchant for not taking things slowly in fear that I'll be dead pretty soon, this suited me perfectly.

Blame it on my neurodivergence but I will never understand why you have to pretend you only kind of like someone for the first couple of months, jump through imaginary made up milestones and then suddenly let the floodgates of external emotions. It's all fake as fuck- decaf, with oat milk and two teaspoons of artificial sweeteners is not how I take my tea.

Alcoholism has been destroying relationships since our common ancestor learned that if you ferment certain substances they make you giddy, carefree and have the sudden urge to text your ex "I miss you". But in the case of Cat and Geordie, it appeared to be what underpinned their relationship.

Their shared love of vodka- which they called potato and their trauma bonding over its mutual destruction over their pasts. Alcohol in a typical relationship is the cruel mistress that tempts one of the two lovers from their bed. Drink was the third member of the throuple Cat and Geordie were in. They'd both wake up every morning, rattling with the shakes from their withdrawal and both pursue their single-handed mission to get hold of some potato.

Generally speaking Cat was always in good form, unless you pissed her off of course and got to witness the near artistic form of her right hook- I would know. But there remained one time in the day where Cat was not to be fucked with for any reason. Every morning when Cat rises she is not Cat, she is the self-titled morning bitch. Geordie would go off on a corner to beg, so he could give the morning bitch her medicine.

They wouldn't argue often but when they did, the trees, the flowers, wheelie bins, seagulls and squirrels would hear it and know about it. All couples argue of course, but when you have no domicile, every argument is a very public domestic. In spite of this, when they said they loved each other, I know they really meant it.

Regardless of how short their period of courting was- their love was genuine and their feelings were real. This was doubled by Cat's belief that you only get three loves in your life, two of her loves had died. As far as she was concerned- this was her last shot at being loved and being in love.

Her determination to be happy made me see that I have no excuse but to demand the same for myself. It makes all the petty things I let great love pass me by seem ridiculous, immature and wasteful. I remembered each love I lost as I lay alone in my sleeping bag looking at the handful of stars sparsely littered across the polluted sussex sky.

One day after they had a catastrophic argument that could have woken a hibernating bear from his slumber, Cat took me to one side. I assumed she wanted some advice about how to deal with the fact that Geordie had been taken away by men in white coats and sectioned before she helped break him out a couple of days later. It was the opinion of myself and the rest of the rest of the bench that she had done this far too soon, but she could not be without him and he could not be without her or drink.

But this was not what she had summoned me to discuss. She wanted to propose to Geordie. I had learned at this point not to give into astonishment. I was legitimately as excited as she was. I had to ask her if he would be okay with being proposed to and if he wouldn't be slightly emasculated by the whole affair. Ever the street spiritualist Cat explained to me that on a leap year women were allowed to propose to men- this was my life now and my new normality, so I nodded and smiled.

When the big day came, Cat tapped me randomly on the shoulder on what up until that point felt like a pretty normal wednesday afternoon on the bench. "It's time," she said. I was a bit shocked she was going to do it on the bench and almost stopped her but this is what she wanted and let's face it…where else would they have done it- I think the Ritz was fully booked that night.

He was sitting on the bench drinking neat vodka from a bottle when she came over to him and got on one knee. No one was watching so I pestered everyone to get their attention. She got down on one knee and said those famous words. In typical Geordie fashion he put his hands on her head and pushed them towards his crotch and in typical Cat fashion she slapped him on the balls and then they were engaged. I hope they agree, but believe I understand why they clung to each other on that long bench.

Their engagement was, like everything else on the bench, a mix of comedy and tragedy, heartfelt and absurd in equal measure. But it was also an emblem of the peculiar resilience of love on the streets. Love here didn't follow the rules or conform to anyone's expectations.

It wasn't about candlelit dinners or whispered sweet nothings. It was about survival, about finding someone who could weather the storm with you, someone who could stand beside you as the world spat in your face and still say, "I've got you."

Cat and Geordie's relationship had a rawness that you rarely see in the polished, curated versions of love society peddles to sell valentines cards and 2 for 1 cinema tickets. Their fights were as intense as their laughter, their loyalty as fierce as their tempers. They were unapologetically themselves with each other, and maybe that's what made them work. On the bench, there's no room for pretense. You can't hide who you are when you're sharing a tent that smells of damp and despair, or when you're splitting a bottle of potato on a Tuesday morning.

But their love, as messy and imperfect as it was, reminded the rest of us that even here, in the margins, humanity persists. Watching them, you couldn't help but feel a little envious. They had something real, something that cut through the grime and the hopelessness. And in their own chaotic way, they brought a bit of light to the bench. They were proof that even in the darkest places, people can still find each other.

Of course, not everyone saw it that way. There were plenty of muttered remarks, plenty of eye rolls and exasperated sighs whenever Cat and Geordie had one of their public rows. "Fucking hell, not again," someone would groan as Geordie's Geordie accent turned into a drunken roar, and Cat's sharp tongue lashed out like a whip. But beneath the grumbles, there was a grudging respect. They were ours, part of our dysfunctional little family, and no matter how much they got under your skin, you couldn't help but root for them.

Their love wasn't perfect. It wasn't healthy. It wasn't the kind of love that inspires rom-coms or Hallmark cards. But it was theirs. And in a world that had taken so much from them, that counted for something. They clung to each other on that bench because they had to, because letting go wasn't an option. And as much as I hated the bench, as much as I hated what it did to people, I couldn't help but admire that.

I left the bench eventually. I clawed my way out, dragging myself back into a world that felt alien and hostile after so long away. But I think about Cat and Geordie often. I wonder if they're still there, still fighting and laughing and holding each other up. I hope they are. I hope they found a way to make it work, a way to hold on to each other in a world that doesn't make it easy. Because if they could, then maybe there's hope for the rest of us. Maybe love, messy and imperfect as it is, really is enough to keep us going.

"Brighton, a place of seedy elegance, where faded grandeur meets the grit of everyday life." —Graham Greene

Stevie G

I have always gravitated towards people who you might describe as *"Fucking Legends"*. It pays to be specific here, not "legends"but "FUCKING LEGENDS". A legend is someone with a full trolley of shopping who lets you go in front of you in the queue at Tesco when all you're buying is a 4-pack of red stripes and a hyperinflated meal deal. I'm talking about the full-blooded, mythic, hell-raising breed. A FUCKING LEGEND is Stevie G.

Every town has one—someone like Stevie, stamped into the collective memory of the place, woven into the local lore with as much grit and permanence as the bricks on the pavements.

A person who is a living memory, a walking story you've probably heard a dozen times, whether you asked for it or not. I came to find that most people had a Stevie G story, he'd just been both a legend and in the same city for long enough to have done something wild in front of you at some point. My

150 days are littered with more of his escapades than I had hot dinners.

Stevie G looked like a caricature, blessed with an appearance you'd be hard-pressed to forget. He was in his mid-40s but strangely looked as if he could be a lot older owing to the few teeth he still had left in his mouth and a full head of silver hair. He was 6ft something and would always sporting the jazziest clothing- my personal favourite being a bucket hat with the word cunt written all over it.

His voice was pure gravel. So raspy it sounded as if someone put a mouthpiece on a 20 pack of Benson & Hedges- distinctive as anything. I called him Uncle Stevie because he reminded me of the Uncle that wouldn't put a tenner in your card because he was saving the money for the high-class escort he would get you on your 18th. *Fucking Legend.*

What kept Stevie young was his god-given athleticism. He could rollerblade, skate, backflip, frontflip and a whole box of party tricks. He made everything seem like a personal stage. His rollerblading is what gave him the most infamy in Brighton. Teaching none other than Katie Price how to skate. Yeah, that Katie Price. Stevie was 17, and so was she—just a local girl with big dreams, none of the celebrity sheen that'd come later. He'd tell this story with pride, his voice dropping to a gravelly whisper, eyes gleaming as he added the punchline: he'd slept with her, before the fame, before the surgeries. *Fucking Legend.*

One Stevie story tops them all. It was Brighton Pride 2024 and I was walking down the New Road. There was a crowd of people standing in front of the bench looking up and laughing at a tree. I couldn't work out what was going on due to my short sightedness. When I got close enough I could see that there was a woman who appeared to be stuck pretty high up the tree.

I was disgusted to see two police officers in the crowd looking up and laughing. I rushed over to and proceeded to give them a piece of my mind and question their inaction. As I did they laughed at me as if they knew something I didn't.

Suddenly, a familiar voice shrieked from the tree "TK, it's pride darling, today I'm not Stevie G, I am Stephanie!". He came down from the tree wearing a skimpy dress, a full face of makeup, and a blonde wig. His feet befriended the floor again for a split second until they re-parted for him to do a signature Stevie G backflip. The crowd erupted in applause and I joined in chanting "stevie g, stevie g is good for me" to the tune of the children's marching song. Because how could you not? *Fucking Legend.*

What I admire most about him is that he defines the meaning: of care-free like no one else on this earth. He told me about spending his teens in what he called:
"*naughty boy schools*" places where they sent the ones who couldn't be tamed.

It was there he learned he grew a disdain for authority figures and a taste for anarchy.He just wanted to have fun and thus decided to devote his life to it. In a world full of order, Stevie G is a glorious, roaring mess—a walking contradiction who laughs in the face of control, the kind of legend you'll talk about long after he's gone.

Brighton's very own *Fucking Legend*, each outrageous story adding to the myth. He makes you wonder if the world could use more people like him, or if it would burn down entirely if it ever tried.Steve was as addicted to adrenaline as any man in Brighton. He enjoyed sticking out like the sorest thumb in the city as much as seagulls like stealing chicken nuggets from unsuspecting tourists.

"In Brighton, you find a unique kind of freedom. It's a town that doesn't judge you, even if it whispers about you behind your back."
—Peter Ackroyd

The Sound of Silence

The streets have a sound to them. It's not the cars or the seagulls or the hum of the city doing its thing; it's something deeper. It's the sound of a hundred people trying to make themselves invisible. A hundred whispers, "Don't look at me," while secretly screaming, "Please, for the love of God, *see me.*"

I didn't notice it at first. The silence creeps up on you. It's the gaps in conversations, the unspoken rules of who gets to speak and when. It's the quiet nods, the shared smirks, the looks that carry entire sentences without a single word spoken. When I first got to the bench, I thought the quiet was indifference. Now I know it's survival. You don't waste words unless they matter, not when you're on the edge of everything.

There's one guy, Gaz. He's the quietest person I've ever met. Not in a meek, shy kind of way, but in a way that makes you feel like every word he doesn't say is waiting for you in the space between breaths. Gaz has this stare, like he's trying to calculate whether you're worth the oxygen it would take to speak to you. Most people don't pass the test.

Gaz doesn't like me much. He doesn't like anyone much, to be fair, but I take it personally because, well, that's who I am. I tried breaking through the wall a few times, starting with casual comments like, "Nice day, isn't it?" or "Where'd you get that jacket?" He'd look at me like I'd just farted in an elevator and then go back to staring at whatever distant, existential hellscape he was fixated on.

One day, I pushed my luck. I asked him what his story was. He lit a roll-up, took one slow drag, and said, "What's your story, mate? Why are *you* here?" The words felt like knives wrapped in velvet. I didn't know what to say, which is rare for me. Gaz had flipped the script, and I didn't have a clever answer. So, I just told the truth: "Curiosity, mostly. Wanted to see what it's like."

Gaz laughed. Not a chuckle or a polite little giggle, but a full-on belly laugh, like I'd told the best joke in the world. "You're mental," he said between gasps. "Absolutely fucking mental." And just like that, the ice was broken. Sort of.

Turns out, Gaz doesn't talk much because he doesn't need to. He watches, listens, absorbs. He's the kind of guy who knows more about you than you do after five minutes of sitting next to you. He pointed out things about me I didn't even know I was doing: how I fidget with my hands when I'm bored, how I stand with my shoulders slightly hunched when I'm trying to blend in. It was unnerving and weirdly comforting at the same time, like being seen for the first time in ages.

Gaz has his own sound, though. It's not his voice—it's the scraping of his lighter, the rustle of his jacket, the way he taps his foot when he's thinking. It's the noise he makes when he inhales his roll-ups, this little crackle that feels louder than it should. It's the absence of the usual chaos, the way everything around him seems to quiet down like the world's holding its breath.

One night, I asked him why he doesn't talk more. He looked at me like I'd just asked why the sky is blue. "Words are expensive," he said. "Why waste them?" That stuck with me. Gaz has a way of saying things that feel like riddles, like there's some deeper meaning you're supposed to figure out. Or maybe there's no meaning at all, and he's just fucking with you. Either way, it works.

The sound of the streets isn't always silence. Sometimes it's shouting, singing, laughing. But the quiet moments are the ones that hit hardest. It's in those moments you realise how much noise we use to distract ourselves. Take the noise away, and what's left? Just you, your thoughts, and whatever truth you've been trying to avoid.

For Gaz, the silence isn't avoidance. It's armor. It's a weapon. And maybe, just maybe, it's the only thing keeping him together. I still don't know much about him. I don't know where he's from or what led him here. But I know his silence, and I know what it means. It's not indifference. It's survival. And on the bench, survival is everything.

The streets don't sing; they hum, they groan, they spit. Their soundtrack is raw, unpolished, the kind of music you don't choose but can't help but hear. It's not beautiful, but it's honest—relentless and indifferent, like the world itself. The streets remind you every second where you are and who you're not.

Mornings start soft, deceptively hopeful. The hiss of coffee machines mixing with the dull thud of shuttered shopfronts groaning awake. Buses groan and grumble in the distance, gears grinding against the day. There's a rhythm to it, sure, but it's not for you. It's for people going somewhere, doing something. For those of us on the bench, the morning chorus is an overture to a play we're not in. The world rolls on, indifferent, and we're just the static in its perfect, polished broadcast.

Then come the voices. Overheard snippets of lives you'll never live. Mothers scolding children about forgotten homework. Couples locked in trivial debates over what's for lunch. A suit shouting into his phone about deadlines and deals, pacing in circles like he owns the pavement.

Their words aren't for you, but they fill the air like stray notes from an orchestra. They don't notice you, and if they do, it's a flicker of disapproval, pity, or discomfort quickly swallowed by their own routines.

By noon, the city throbs with energy. Footsteps, endless footsteps, each telling its own silent story. The click of heels, the shuffle of sneakers, the thud of boots worn down by miles and months. It's an unintentional rhythm section for the day, underscored by the jangle of coins in pockets and the rustle of fast-food bags. Occasionally, laughter punctuates the noise, a sharp contrast to the monotony, fleeting and untouchable.

But nightfall—the streets take on a different tone. The bassline drops; the volume cranks up. It's not just noise anymore—it's edge. Laughter turns sloppy and guttural, words slurred and distorted by cheap vodka and cheaper decisions. Arguments cut through the air, raw and primal, spilling out of pubs and alleyways. Glass shatters. A car alarm wails. Somewhere, fists meet flesh with a sickening thud, and the resulting groans linger long after the participants vanish into the shadows.

The night is heavy, oppressive. Every sound feels sharper, closer, even when it's not for you. Sirens wail, distant at first, then louder, echoing against the brick and steel of the city like war cries. You freeze for a moment—are they coming for you? No, not this time. A door slams. A drunken laugh cuts through the tension, hollow and haunting.

And then there are the small, almost invisible sounds—the ones you notice only when everything else fades. The rustle of a plastic bag as someone lays claim to their corner for the night. The clinking of bottles as a man counts his drinkable fortune. The shuffle of someone rearranging layers of cardboard to fend off the cold. These are the sounds of survival, a quiet rebellion against the indifference of it all.

By the early hours, the city is exhausted. The sounds slow, soften, scatter. A lone car engine purrs at a red light, the driver staring into nothing. A seagull squawks into the void, its voice bouncing off silent streets. The wind whispers, carrying the briny breath of the sea and the acrid tang of rotting trash. For a moment, the city feels still, almost at peace. Almost.

Out here, the noise never really stops. The streets don't let you forget. Every sound—every stomp, shout, shuffle—is a reminder. Of the chaos. Of the humanity. Of the quiet dignity in just existing. The sound of the streets doesn't belong to anyone, but it owns you.

Whether you want it to or not, you learn to listen, to let it crawl under your skin, to hum along with its discordant tune. It's the only music you get. And after a while, you start to wonder if you'd even want silence.

The Devil's Lettuce

It took God seven days to make the world, but it took the Devil his whole life to create spice. If ever there was a substance tailor-made for the damned, this was it. The fact that it was called a "legal high" gave it a twisted appeal. Why on earth would our dear, benevolent government, who loves us so fucking much, invent a drug and make it legal unless it was safe as... fuck that—safer than houses? In short, what could possibly go wrong?

The answer, as it turned out, was everything.

Spice is what you get when humanity loses the plot. It's not a drug; it's a weapon, a chemical Frankenstein designed to scramble brains and hollow out souls. You don't take spice; it *takes* you. I first came across it as a teenager, and even then, the allure was impossible to miss. The promise of a high that wouldn't land you in jail? It sounded too good to be true. And, of course, it was.

For most of the bench, spice wasn't a choice. It was a consequence, a ghost that followed them out of prison. It's the perfect prison drug, really. Cheap, easy to smuggle, and the high? Well, that's the kicker. It doesn't make you feel alive; it makes you feel *nothing*. It's not an escape—it's a temporary death. For men and women staring down years behind bars, that nothingness is seductive. The idea of sleeping through a sentence, of turning years into a blur, is a comfort. Spice doesn't just dull the edges of reality; it erases them.

But out here, on the streets, spice doesn't act as an escape—it's a leash. It turns people into zombies, shambling, broken shells of who they were. I've seen it up close, the way it drags people down. I once watched a guy on the bench take a single puff, then collapse like a marionette with its strings cut. His body contorted in slow, horrifying spasms while the rest of us stood around, unsure whether to help or run. He came back to himself after a few minutes, looked at us through glassy, lifeless eyes, and smiled. "Good batch," he muttered. Good? That was the worst thing I'd ever seen. But to him, it was just another day.

Then there was Cameron and Stacey. If spice had a mascot, it would have been those two. Cameron was wiry, jittery, and always seemed like he was halfway through an argument with himself. Stacey, on the other hand, had this weird kind of grace to her, even when she was falling apart. They were inseparable in the way only addicts can be, bonded by their shared misery and their mutual need for the next hit.

Stacey had been beautiful once; you could see it in the remnants of her face, the high cheekbones, the eyes that still held a spark of mischief even when she was high. I used to joke and call her a "Spice Girl." It was cruel, I know, but humor is sometimes all you have out there. She took it in stride, though, laughing in that raspy, broken way of hers. "Spice World, innit, TK?" she'd say, waving her arm around the bench like it was the set of a music video.

Cameron wasn't as fun. Spice had eaten away at whatever charm or personality he'd once had, leaving him hollow and unpredictable. He'd go from laughing to screaming in seconds, his mood swinging wildly with no warning. Stacey calmed him down, mostly. She had a way of bringing him back from the edge, holding his face in her hands and whispering to him until his breathing slowed. But when they were both high, it was like watching two ghosts try to hold each other.

One night, I found them slumped against each other on the bench, barely conscious. Stacey's hand was still clutching the baggie, her fingers curled around it like it was a lifeline.

Cameron's head lolled to the side, his eyes half-open but unseeing. It was a quiet moment, almost peaceful, but it wasn't real. Nothing about spice is real. It's all synthetic, from the chemicals to the calm it promises. It takes and takes until there's nothing left to give.

I don't know what happened to Cameron and Stacey. One day they were there, the next, gone. That's how it is on the bench. People disappear without warning, swallowed up by the streets or the drugs or the system, and you never find out where they went. But I'll always remember them, the way they clung to each other even as the spice tried to tear them apart. In their own twisted, tragic way, they loved each other. Or maybe they just needed each other. Sometimes it's hard to tell the difference.

Spice isn't just a drug. It's a predator, a shadow that creeps in and devours everything good and human. It promises escape but delivers chains. And yet, for the people on the bench, for the lost souls like Cameron and Stacey, it's a comfort, a way to make the unbearable just a little more bearable. It's hell in a baggie, and it's everywhere.

Spice is insidious, not because it's seductive, but because it's efficient. It's the drug of desperation, tailor-made for those who have nothing left to lose. You don't see people walking into a life of spice; they stumble into it, usually already broken, already dangling on the edge of something worse. For the bench, it's the final nail in a coffin they've been building their entire lives, plank by plank, hit by hit.

What makes spice so horrifying is how it preys on vulnerability. For those just out of prison, it's the thing that numbs the shock of reentry, of being tossed back into a world that no longer makes sense. For those on the streets, it's the cheap high that drowns out the hunger, the cold, and the crushing weight of failure. It doesn't make anything better—it just makes everything stop for a while. And sometimes, that's enough.

I've seen the spice zombies, the walking dead shuffling through the alleys of Brighton. They move in jerky, unnatural motions, their bodies puppeteered by some unseen force. Their eyes are empty, their faces slack. It's not like being drunk, or even like the sluggishness of heroin. It's more alien than that. Spice doesn't just mess with the body—it rewires the brain, turning people into hollow shells, strangers to themselves.

There's this spot near the pier where you can always find them, sitting or lying in heaps, their limbs twisted into awkward angles, their gazes fixed on nothing. Tourists pass by, horrified but fascinated, like they're watching some macabre street performance.

Some take pictures, their phones clicking away as though they're capturing a moment of culture rather than someone's worst day. Others clutch their children's hands a little tighter and mutter about the state of society. But none of them stop to help. What would they even do if they did?

Spice has a way of erasing humanity. It reduces people to statistics, to shadows, to nuisances. It makes it easier for the rest of the world to look away. "They did it to themselves," people say, as though that absolves them of any responsibility, any obligation to care. But no one chooses spice. No one wakes up one day and decides to destroy their life for the hell of it. Spice isn't a choice; it's a symptom. A symptom of poverty, of trauma, of a society that's more comfortable blaming individuals than fixing the systems that broke them.

And yet, even amidst the destruction, there are moments of clarity, glimmers of the people who used to be. I remember one guy, a regular on the bench, who was deep into spice but still had a knack for poetry. He'd sit there, his body trembling, his voice slurred, and recite these haunting verses about life and loss and everything in between.

He called himself an "urban bard," and for all his faults, he had a way with words. But spice was stealing him, one poem at a time. Eventually, he stopped reciting altogether, his voice swallowed up by the haze.

There's no easy solution to the spice epidemic, no magic wand that will fix it overnight. But there are steps we could take, if we cared enough to try. Harm reduction programs, safe consumption sites, education campaigns that actually resonate with people instead of just wagging a finger at them—these are all pieces of the puzzle. But the real work lies in addressing the roots of the problem: the poverty, the lack of affordable housing, the gaping holes in mental health care. Until we tackle those issues, spice will keep spreading, devouring more lives, more families, more futures.

The bench is full of people like Cameron and Stacey, people who have been chewed up and spat out by a world that never gave them a fair shot. They cling to each other, to their baggies, to whatever scraps of hope they can find. And while it's easy to judge them, to write them off as lost causes, the truth is they're just like anyone else. They laugh, they cry, they fall in love. They dream of better days, even if they don't believe they'll ever come.

Spice is a thief, but it hasn't stolen everything. Not yet. There's still humanity on the bench, still people worth saving, still lives that could be rebuilt if we had the will to do it. The question isn't whether they're worth saving—it's whether we're willing to admit that we've failed them and do something about it. Because until we do, the bench will remain a graveyard of potential, a place where hope goes to die. And spice will keep winning.

Dignity

Dignity is a funny thing. It's invisible, intangible, but when you lose it, it's like losing a limb. You feel the absence of it in everything you do. On the streets, dignity is a currency as valuable as any loose change in your pocket, and just as fleeting. It's there one moment, gone the next, and most of the time, you don't even realise it's missing until it's too late.

I used to think dignity was something you carried within yourself, like confidence or pride. But out here, you learn quickly that dignity is more like a fragile shell—something that the world can strip away from you piece by piece. It's in the way people look at you, or don't look at you. It's in the stares of disgust, the averted eyes, the way someone walking past will hold their breath as if poverty were contagious. It's in the tone of voice when someone says, "Sorry, mate, no change," like they're apologising for your existence.

Take Benny, for instance. Benny was one of the old-timers on the bench. He had this quiet dignity about him, a kind of unspoken pride that seemed out of place among the chaos. He wore a threadbare suit jacket every day, no matter the weather, and he always kept his boots polished. They were falling apart, but they shone like mirrors.

Benny wasn't a beggar; he was a performer. He'd sit on the pavement with his harmonica, playing old blues tunes that sounded like they belonged to another century. People would stop and listen, some would toss coins into his hat, but most just walked on by. Benny didn't care. He played for himself, for the music, for the little piece of his soul that was still his own.

One day, Benny's harmonica disappeared. Stolen, probably. On the streets, everything is up for grabs if you're not holding it tight enough. Benny sat on the bench that day, staring at his hands like they'd betrayed him. Without the harmonica, he was just another homeless man, invisible, irrelevant. His dignity had been taken from him, and there was nothing he could do to get it back.

Then there's Rita. Rita's dignity had been stripped away long before I met her. She was middle-aged, though the streets had aged her beyond recognition. Her voice was raw from shouting at passersby, demanding to be noticed. She'd rant and rave, cursing the world, the government, the weather, anyone and anything that crossed her mind. Most people avoided her like the plague, but I liked Rita. She had fire. Even when she was at her worst, there was something fierce about her, something defiant.

One night, I found Rita crying behind a bin. It was the first time I'd ever seen her vulnerable. She told me that a group of kids had thrown a milkshake at her earlier that day, laughing as it splattered across her face and clothes. "I didn't even say anything to them," she whispered, her voice cracking.

"They just… did it." I sat with her for a while, not saying much, because what could I say? There's no way to fix something like that. The milkshake was gone, but the shame lingered, etched into her skin like a scar.

Dignity isn't something you lose just because the world decides you don't deserve it. It's not taken from you by circumstance, by empty pockets, or by the cruel stares of people who don't know better. Out there, on the streets, dignity is a choice—one you make every day, no matter how much the odds are stacked against you.

"In Brighton, it's impossible to get away from the contrasts—the opulence of the Regency architecture against the rawness of the sea and sky." —Virginia Woolf

Bottling it Up

This may sound somewhat controversial but, alcohol on the streets isn't a luxury—it's a lifeline. It's not just something to take the edge off; it's the edge itself, sharp and unforgiving, wrapped in a thin, comforting haze. Out there, the bottle is a constant companion, a therapist, a lover, and an executioner all rolled into one.

For most people on the bench, drinking isn't a choice; it's a necessity. It numbs the cold, drowns the hunger, and quiets the noise of a world that's always telling you you're not enough.

I'd be lying if I said I didn't need alcohol to get through a day out there. Not want—need. Out there, it was less a vice and more a necessity, like shoes or a half-decent coat. There's no way in hell you could sit back and watch all the bedlam without the rose-tinted glasses of drink.

Without it, every insult hurled, every fist swung, every tear shed, and every injustice endured would hit you like a freight train, leaving you raw, exposed, and utterly powerless. The drink dulled the edges of reality, just enough to make it survivable. You didn't drink to celebrate or socialise or even to drown your sorrows. You drank to stay sane, to blur the edges of the world enough that it didn't feel like it was always on the verge of swallowing you whole.

Alcohol became both a friend and an enemy, a double-edged sword you couldn't let go of. It wasn't just about drowning sorrows or numbing pain—it was a coping mechanism, a social currency, and sometimes, a lifeline. A shared bottle was the universal icebreaker, the olive branch that turned strangers into allies. It was a toast to making it through another night and a distraction from the endless monotony of surviving without a home.

It wasn't glamorous. It wasn't romantic. It was functional. Vodka, cider, whatever came to hand. No one on the bench was sipping single malt and discussing the finer points of oak-barrel aging. This was the realm of two-liter bottles, flimsy cans, and liquids that smelled like regret. Drinks that burned your throat on the way down and your dignity on the way out. But that didn't matter. What mattered was that they worked.

I watched as it consumed the people around me. The mornings always started the same: rattling hands, shaky voices, eyes darting like they were searching for a way out of their own bodies. Withdrawal was an unforgiving master, and the only way to silence it was to feed the beast.

I'd see them scrape together coins, beg, borrow, and barter to get their first drink of the day. And once they had it, the relief was almost religious. The shakes would stop, the world would soften, and for a brief moment, they were themselves again.

The first swig was the best. That brief moment when the warmth spread through your chest and softened the edges of reality. After that, it was just maintenance—keeping the buzz alive long enough to get through the day, or the night, or the next conversation with the guy who hadn't seen his kids in twenty years but wanted to tell you his life story. Of course, there were casualties.

There always are. Alcohol isn't a team player. It takes what it wants, and it doesn't care who gets hurt in the process. It was like watching a battle from the sidelines, knowing full well you could be next. Cat, for all her fire and ferocity, was a different person before her first drink of the day. "The morning bitch," she called herself. Until she had her medicine, she was a time bomb, ticking and ready to blow. Then there was Geordie, who treated vodka like a sacrament.

Cider was the drink of choice for most. Cheap, strong, and readily available, it was the fuel that kept the bench alive. "Scrumpy", though there was nothing rustic or charming about the industrial-grade poison sold in plastic bottles for pocket change.

I remember once watching a guy polish off an entire bottle in one go, tipping his head back like it was water and he'd just crawled out of the Sahara. He wiped his mouth, let out a satisfied sigh, and said, "Breakfast is served." I laughed, but it wasn't funny. Not really.

The thing about alcohol is that it's both the problem and the solution. It creates chaos, and then it quiets it. Arguments would erupt over the smallest things—a misplaced can, a careless comment, a glance that lasted too long. But as the bottles emptied, the anger would fade, replaced by a dull camaraderie. By evening, the bench would be a chorus of slurred laughter, mumbled stories, and off-key singing. For a little while, the world didn't seem so bad.

There was one guy, Danny, who lived for the drink. He was always the first to crack open a bottle in the morning and the last to pass out at night. Danny had a knack for making you laugh, even when you didn't want to.

He'd tell these ridiculous, over-the-top stories about his glory days, most of which were clearly bullshit, but you didn't care because the way he told them was pure gold. He'd gesture wildly, acting out the parts, his voice booming across the bench like he was on stage at the West End.

But when Danny wasn't drinking, he was a different person. Quiet, withdrawn, almost unrecognisable. I asked him once why he drank so much, and he looked at me like I'd just asked why the sun rises in the east. "Because it stops me thinking, mate," he said. "It stops the bad stuff from coming back." He didn't elaborate, and I didn't press. You learn quickly not to dig too deep out there. Everyone has their demons, and most of them don't want to be exorcised.

But the price of alcohol is steep, and not just in pounds and pence. I've seen it destroy friendships, health, and lives. I've watched people lose themselves to it, their personalities eroded sip by sip until nothing was left but the thirst. And yet, I understand its pull. Out here, where every day feels like a battle you can't win, the bottle offers an escape, however fleeting. It's not a solution, but sometimes it's the only thing that makes the pain bearable.

I tried to keep my distance from the edge, but it was a constant battle. Alcohol had a way of luring you in, whispering promises of peace, of numbness, of just one more hour without thinking about where you were and how you got there. It was tempting, especially on the nights when the cold bit through your coat and the loneliness gnawed at your insides. On those nights, a drink wasn't just a drink—it was survival.

But survival came at a price. The drink didn't solve anything. It didn't make the nights warmer or the mornings brighter. It didn't fix the loneliness or the fear or the ache in your chest that never quite went away. It just dulled it for a while, like pressing a cloth over a wound that wouldn't stop bleeding. And when the buzz wore off, the world came back sharper and meaner than ever, and you were left with nothing but an empty bottle and a headache that felt like penance.

Alcohol was the great equaliser on the streets. It didn't matter if you were fresh to the bench or a grizzled veteran of its particular hellscape; it had a way of leveling everyone out. For some, it was a crutch, for others, a coping mechanism.

For a few, it was a full-blown religion—bottles their holy relics, the corner shop their chapel. But for everyone, it was constant. It was the thread running through the days and nights, stitching together moments that might otherwise have been unbearable.

The way people drank told you a lot about them. There were the careful drinkers, the ones who sipped slowly, rationing their poison like they were preparing for a long campaign. Then there were the binge drinkers, the ones who poured it down their throats as if trying to outrun whatever demons were chasing them.

And then there were the functional drinkers, the ones who somehow managed to stay upright, coherent, even charming, despite the fact that they were rarely, if ever, sober. These were the ones who fascinated me the most, the ones who made you wonder how much of their personality was real and how much was just the drink talking.

Alcohol also had a way of turning the mundane into the absurd. I remember watching a guy, must've been in his fifties, sitting cross-legged on the pavement with a bottle of Frosty Jack's. He wasn't just drinking—he was performing. He'd take a swig, then launch into a dramatic soliloquy about the decline of British football or the meaning of life or why pigeons are secretly government spies.

It was nonsense, of course, but it was captivating nonsense. And the thing is, he believed it, at least in that moment. The drink had given him a stage, and he was going to make the most of it.

But the flip side of alcohol's theater was its violence. The same bottle that could make someone a poet could also turn them into a monster. Fights over who drank the last drop, arguments that escalated into full-blown brawls, the unpredictable rage of someone who'd had just enough to tip them over the edge. It was always there, lurking beneath the surface, ready to explode.

And when it did, it wasn't just fists and broken bottles flying towards broken bodies—it was words, too. Words that cut deeper than any punch, words you couldn't take back once they'd been said. Out here, a drunk man's tongue was sharper than any blade.

I watched it destroy people. Watched it turn fire into ash, hope into resignation. I saw it strip away the last shreds of dignity from men who had nothing else to lose. But I also saw how it held them together, how it kept them going when nothing else could.

Out there, the drink was a paradox—a poison that kept you alive. A cruel kind of salvation. And whether you liked it or not, you learned to make peace with that. Because out there, you didn't have much choice.

Policing the Homeless

I don't buy into the "All Cops Are Bastards" rhetoric. In my former life, I had never had any trouble with the police that I didn't deserve. I have an unbelievable amount of respect for Essex police who have always treated me with courtesy and respect.

They are steady hands, just another part of the landscape. They are like the steel beams of the Dartford Crossing or the endless rows of council flats. They never gave me that familiar lump in the throat, the lump that comes standard when you're black and staring down a badge. They'd watch me, sure, but they didn't watch me like a fox in a hunt. They never quite hit that nerve. They never crossed that invisible thin blue line.

Now, the Met Police—that's a different beast altogether. Their reputation precedes them, and it's ugly as sin. It's the kind of reputation that makes people lower their voices, even when they're already whispering. Everyone knows their dirty laundry, well-documented like the kind of scandals that make headlines: corruption, brutality, racism—old scandals that cling to their boots like muddy dog shit.

The stuff of national shame, whispered through the damp alleys and forgotten estates of London. But I never had to deal with that. I could almost sympathise with them—those poor saps pulling double shifts trying to wrangle the nastiest villains in the country for the kind of paycheck that wouldn't cover a good therapist.

But then there's Sussex Police. Or more specifically, Brighton Police. If that had been my only interaction with the boys in blue, I'd have painted the whole force with the same brush, a thick coat of disdain. They harassed me, bullied me, actively hunted me for sport like they needed the exercise, and locked me away on several occasions.

Hatred is a strong word but so is *criminal*. I don't waste my vitriol on the everyday rank-and-file police officers who would be the ones who would arrest me, drive me to the station, book me in, give me unlimited pot noodles and all-day breakfast, drive me back to Brighton from the station when their asshole boss refused to give me a bus ticket to make the 5-mile journey home.

My disdain is reserved for those at the very bottom, the Police Community Support Officers (PCSOs), and those at the very top. Let's start with the bottom, shall we? PCSOs. In their defence, they don't have anywhere near enough power so I don't necessarily blame them for the majority of their wrongdoings but still harbour deep resentment for them and how they do their jobs.

They do not have the power to make arrests so they are essentially professional snitches. If life was school, Police Officers would-be teachers with the power to give you detention, call your Mum, suspend you, recommend your exclusion to the headmaster and beat you with a cane in the 70s. PCSOs on the other hand are the brown-nosing prefects that can give you lines for not having your shirt tucked or your top button undone.

I don't have a negative opinion of every PCSO—I met plenty who were decent enough, just doing their jobs without the swagger or malice that some of their colleagues wield like a badge of honor. No, my disdain is reserved for three particular PCSOs: two young men and a young woman who seemed to treat their uniforms as invitations to bully and belittle. The rest I encountered were fine—courteous even—but those three had a way of reminding you that power, no matter how limited, can always be abused.

Now for their bosses, those at the very top who set the agenda and remind them of what's important (to them at least), who should be arrested, who should be left alone and what to focus on. My view of them can only be summarised in one sentence: they are shit at community policing. No, their game is all about keeping the mess hidden—stuffing the unsightly bits into the shadows where the tourists won't see.

They've got no patience for the homeless, no interest in understanding the human wreckage littered on their doorsteps. Let the homeless drink in peace. I'm damn near certain I wouldn't have any or much of a problem with Sussex Police if I was never part of the bench.

My chief objection to the police is how they give out Community Protection Notices like they were sweets. If you don't know what a CPN is, the best way to think about it is as a legally sanctioned "shut-up-and-get-lost" card, handed out at the discretion of officers who've decided that your existence is an inconvenience to their version of order. It's not a proper conviction or a ticket, but it carries the same sting—a formal declaration that you, specifically, are a problem.

On paper, it's meant to tackle anti-social behavior, but in practice, it's wielded like a hammer against the homeless, a way to sweep the messiness of poverty under the nearest rug.

A CPN can bar you from sitting on a bench, from entering a certain street, or from being in the vicinity of a public building. If you're unlucky enough to get one, it essentially criminalises your life. Imagine being told that you can't exist in the only places left to you.

Now multiply that by the number of times an officer feels like flexing their authority in a given week. The system thrives on paperwork, and CPNs are their favorite kind—easy to fill out, impossible to fight, and useful for getting you out of sight and out of mind.

Take Brighton Pavilion. It's the heart of the tourist circuit, all postcard-perfect Regency charm, but for the bench's inhabitants, it's also the first place you'll get slapped with a CPN. You're not allowed to sit too long or look too rough. The tourists can't be subjected to the sight of poverty; it disrupts the aesthetic. There's a cruel irony in being homeless in a town that markets itself on being free-spirited and bohemian. Brighton loves its quirks, but only the ones that fit neatly on an Instagram feed.

I've seen people break down over those notices. Not because they didn't expect them, but because they knew exactly what they meant. When you're homeless, you're already living on the knife's edge, and a CPN can push you right off it. Lose your spot on the bench, and suddenly, you're in unfamiliar territory—unwelcome in the places you know, unwelcome everywhere else. It's not just inconvenient; it's dangerous.

There's an illusion that the police care about rehabilitation, that these notices are meant to nudge you toward help or housing. That illusion shatters quickly. The officers who handed them out never offered solutions, just ultimatums.

Move along or face arrest. Disappear or deal with the consequences. And when you're dealing with survival on a day-to-day basis, compliance isn't always an option.

The worst part is how CPNs are treated like a panacea for homelessness. They don't address addiction, mental health, or the lack of affordable housing. They don't stop the cycle of arrests, fines, and further disenfranchisement. All they do is shift the problem somewhere else, to another bench, another corner, another town. It's like trying to clean a spill by moving the puddle around with your foot.

Of course, it's not just the CPNs. They're merely a symptom of a broader sickness—the institutional laziness that treats homelessness as a crime instead of a crisis. The officers handing them out aren't solving anything; they're just creating the illusion of action. It's bureaucracy at its finest: hollow gestures wrapped in red tape.

I know this might come across as bitter, and maybe it is. But bitterness isn't unwarranted when you've seen the same faces pushed out of one space after another, herded like cattle with nowhere to go. Bitterness is earned when you watch people being told they don't belong anywhere, over and over, until they start to believe it.

The truth is, the police are just one part of a machine that's designed to manage, not to solve. The real crime isn't sitting on a bench too long or being in the wrong place at the wrong time. The real crime is that we've built a system that punishes people for not fitting neatly into it. And CPNs? They're just another tool in a box full of blunt instruments, wielded without care for the people they're breaking.

The problem with how the police interact with the homeless isn't just about enforcement—it's about philosophy. The entire approach is reactive, punitive, and ultimately designed to push people out of sight rather than address the underlying issues.

If the police and the systems they serve are serious about tackling homelessness, addiction, and street-level poverty, the solutions need to be proactive, compassionate, and rooted in the idea that people on the margins are still people.

Real community policing would mean officers aren't just enforcers—they're connectors, mediators, and advocates. Instead of issuing Community Protection Notices like parking tickets, officers should work alongside social workers and outreach teams to find actual solutions.

That means training officers in trauma-informed care and making empathy as much a part of their job as enforcement. Imagine a system where a homeless person being moved along from a bench comes with a referral to a proper resource—a shelter, a mental health clinic, or a housing service. Officers could carry cards with direct contact information for local organisations, cutting through red tape instead of adding to it. It's not about coddling; it's about giving people a fighting chance.

Police Community Support Officers (PCSOs) are often viewed as powerless and irrelevant—professional snitches, as I've seen them called. But what if their role was reframed? Instead of acting as the first line of defense for gentrification, PCSOs could be redeployed as full-time liaisons for the homeless community.

They could receive specialised training in addiction, mental health, and social services, becoming a bridge between people in crisis and the help they need. Imagine PCSOs working with outreach teams to deliver harm reduction supplies, schedule medical appointments, or even provide transportation to shelters. Instead of being the bad guys, they could become trusted figures—people who show up to help rather than to judge or punish.

The Community Protection Notice system, as it exists, is little more than a thinly veiled mechanism for criminalising homelessness. If the goal is truly to address anti-social behavior, CPNs need to be radically rethought—or better yet, abolished. The focus should be on restorative practices that deal with behavior, not punitive measures that criminalise existence.

For example, if someone is repeatedly causing disturbances due to addiction, instead of banning them from public spaces, mandate participation in a harm reduction program. Pair this with resources—transportation, case workers, housing support—to make sure they have what they need to actually comply. The goal should be to stabilise, not to alienate further.

The evidence is overwhelming: Housing First works. Give people stable housing without preconditions, and the rest—sobriety, employment, mental health—has a chance to follow. Police and councils could collaborate to identify individuals who are repeatedly targeted by enforcement and prioritise them for Housing First programs.

When people aren't constantly moving from one doorway to the next or facing arrests for existing in public, they can start to rebuild their lives. The police would still play a role here, but a supportive one. Instead of seeing the homeless as nuisances, officers could act as advocates within the system, pushing for people to be prioritised for housing and services.

One of the biggest failures of modern policing is the lack of understanding about trauma. Many people on the streets have experienced unspeakable things—abuse, violence, neglect—and their behaviors often stem from this pain. Police need to be trained to recognise and respond to trauma rather than escalating it.

This doesn't mean excusing violence or criminal acts, but it does mean approaching situations with an eye toward de-escalation rather than dominance. Imagine if, instead of shouting orders and threatening arrest, officers could approach someone in crisis with calm, empathetic language. De-escalation isn't just about avoiding violence—it's about building trust, about showing people that not every interaction with authority has to end in conflict.

Policing homelessness is a losing game if it's done in isolation. The police need to work in lockstep with social services, housing authorities, and nonprofits. This could mean embedding social workers into policing teams or creating joint task forces specifically for addressing homelessness.

For example, when an officer encounters someone living rough, a social worker could be called in immediately to assess their needs. Are they struggling with addiction? Mental health? Do they just need a safe place to sleep? The response could be tailored rather than one-size-fits-all.

Finally, the broader societal attitude toward homelessness needs to change. The reason the police focus so heavily on pushing the homeless out of visible areas like Brighton Pavilion is because society doesn't want to see its failures. The best way to address homelessness is to stop pretending it's an individual problem and start acknowledging it as a systemic one.

These solutions aren't radical—they're practical, and they've been proven to work in cities around the world. What's radical is the idea that we could look at them as people worth saving. Until we do that, the cycle will continue, and the bench will keep filling with people who deserve more than this world has given them. It's not impossible to change—it just takes the will to do it.

"Brighton: a town that dances on the fine line between light and dark, always teetering but never falling."—Patrick Hamilton

The Longest Nights

Nighttime on the streets isn't just a part of the day; it's a world of its own. When the sun dips below the horizon and the last shops lock their doors, the city changes. It's not just quieter; it's darker in every sense of the word. The streets, so bustling and alive during the day, become a shadowy battleground where survival instincts kick into overdrive. The nights are cold, long, and unrelenting, and the only thing you can count on is that the darkness won't care whether you make it through or not.

For most of us on the bench, night meant finding safety, and safety was a delicate balance. Too visible, and you became a target for drunks or bored teenagers looking to stir up trouble.

Too hidden, and you risked worse—being robbed, assaulted, or worse by people who thrived in the anonymity of the shadows. My strategy was always to find somewhere in between. Tucked away, but not isolated. Near people, but not too close. It's a tightrope, and every night you walk it, hoping the wind doesn't blow too hard.

Nights are long when you have nothing to do but exist. Time stretches out in strange ways. Minutes feel like hours, and hours feel like days. You lie there, listening to the city breathe around you, every sound amplified by the silence. Footsteps crunching on gravel, the distant hum of traffic, a bottle smashing somewhere far off. The world doesn't stop moving, but at night, it feels like it's moving without you.

The cold is the first thing to hit you. It seeps into your bones, no matter how many layers you wear, no matter how tightly you wrap yourself up. It's not the sharp, biting cold of winter mornings; it's a slow, creeping chill that wraps around you like a snake and doesn't let go. You can't fight it. You can only endure it.

Sleep, when it comes, is fleeting and restless. Every sound jolts you awake—an argument in the distance, the scuffle of shoes too close for comfort, the rustle of a plastic bag that might mean someone's about to rummage through your stuff. Most nights, I didn't dream. The bench doesn't leave room for dreams. The nights strip you of that luxury, leaving you with nothing but half-formed thoughts and a body too tired to move but too scared to rest.

Not everyone sees the night as something to endure, though. For some, it's when they come alive. The scavengers, the wanderers, the ones who can't sit still. You'd see them pacing the streets, their faces lit by the glow of a cigarette or the faint blue light of their phone screens. They're out there searching—for food, for warmth, for company, for meaning. Maybe all of the above.

The arguments were louder at night. The fights more vicious. Maybe it's the cover of darkness, the way it lets people unleash what they hold back during the day. Or maybe it's just the alcohol, the spice, the sheer weight of being alive in a world that doesn't care if you're breathing. I'd watch from my spot as voices escalated, fists flew, and chaos erupted over something as small as a misplaced can of cider or a careless comment.

But it wasn't all bad. Some nights had a strange kind of magic to them, moments that reminded you there was still light in the darkness. Like the time Benny, the old-timer with the harmonica, played a song so hauntingly beautiful it silenced the entire bench. Or the night Cat shared a bottle of vodka with me and told me stories about her childhood, her voice soft and nostalgic, her eyes distant but warm. Those nights didn't happen often, but when they did, they stayed with you.

Then there was the storm. The kind of rain that doesn't just fall—it attacks. Sheets of water cascading from the sky, soaking everything in their path. My sleeping bag was no match for it, and within minutes, I was drenched. I tried to find shelter, but every doorway and overhang was already taken.

The city felt like it was drowning, and so was I. Eventually, I ended up under a railway bridge, huddled with a group of strangers who looked as miserable as I felt. No one spoke. We just sat there, shivering, waiting for it to end.

And then there were the stars. On rare clear nights, when the pollution and clouds took a break, the sky would open up, revealing a universe so vast it made everything else seem small. I remember lying on my back one night, tracing faint constellations I couldn't name, feeling both insignificant and infinite at the same time. Out there, under the stars, the world felt almost manageable. Almost.

The nights could be cruel, but they weren't always unkind. There were moments of connection, of camaraderie, of shared survival. Like the time we pooled our last few coins to buy a cheap bottle of whiskey and passed it around under a broken bus shelter, laughing and telling stories as the rain poured down. Or the night I met Rob, a conspiracy theorist with wild eyes and wilder ideas, who swore he had proof the Queen was a lizard but wouldn't show me because "they" were watching.

But for every night of laughter, there were ten of silence. And it's the silence that stays with you. The sound of the city settling into sleep, the quiet hum of your own breath, the weight of your thoughts pressing down on you like the cold. Silence is the hardest thing to survive. It forces you to confront everything you've been running from—the regrets, the mistakes, the what-ifs. It strips you bare, leaves you exposed, and dares you to keep going.

Morning always came, though. The first light of dawn creeping over the horizon, the city waking up, the streets transforming. The bench would go from a place of refuge to a stage, where the world would judge you with a glance, a sneer, or worse, indifference. But morning also brought relief. Another night survived, another small victory in a war with no end.

The longest nights taught me more about myself than any day ever could. They taught me what it means to endure, to hold onto hope when the darkness feels endless, to find light in the smallest of moments. They taught me that survival isn't just about making it through—it's about holding onto the parts of yourself that matter. And every time the sun rose, I knew I'd won, at least for another day.

Shadow People

Invisibility might be a survival skill on the streets, but it's one I've never mastered. It's not in my nature. I'm big, bold, and brass—a natural-born extrovert with a voice that carries and a presence that demands attention. I like to be seen, to make my mark, to remind the world I exist. But out here, that's not an advantage. In fact, it's a liability.

The streets reward the quiet, the unassuming, the ones who know how to melt into the background. Blending in is how you avoid trouble, how you keep your head down and stay off the radar of people who'd rather see you gone. But I've never been good at blending in. I stand tall, I speak loud, and I walk like I own the pavement beneath my feet. It's who I am, and I can't turn it off—not even when I probably should.

When I first landed on the bench, I didn't think much about how I carried myself. I'd sit there, reading a book or sipping tea, letting my laughter echo down the street if someone told a good joke. But the stares started coming, and not the kind I'm used to. These weren't admiring glances or curious looks. They were judgmental, suspicious, or outright hostile. Some people even crossed the street to avoid me, as if confidence were contagious and they might catch it.

Being visible out here makes you a target. People notice you, and not always in a good way. Drunks looking to pick a fight, PCSOs on a power trip, or just the everyday passerby who thinks it's their business to tell you how to live your life—they all hone in on the one person who isn't shrinking into the shadows. And that person is almost always me.

Even when the attention isn't hostile, it's rarely welcome. There was a man who stopped me on the street once, a clipboard in hand and a well-rehearsed spiel about helping the homeless. He clearly thought I needed his help—because of how I looked or where I was standing, I don't know. "Mate, I'm not homeless," I lied, just to get him to leave me alone. His face fell, confused. "Oh," he muttered, before quickly moving on to someone else. I wasn't in the mood to be someone's charity project that day. Most days, I'm not.

But for all the trouble it brings, being seen isn't all bad. There's power in refusing to shrink, in standing tall and letting the world know you're still here. I might not blend in, but I stand out for a reason. People remember me. They notice the clean clothes, the book in my hand, the way I carry myself like I'm worth something—even when the world tells me I'm not.

That visibility, that boldness, has its perks. Like the time a woman stopped me outside a café, apologising profusely for staring. "I just wanted to say you look really put together," she said, handing me a coffee. "I admire that." It wasn't much, but it was something. A moment of recognition, of respect, of seeing me as a person instead of a problem.

I know I'm not like most people on the bench. I don't blend in. I don't disappear into the cracks. And while that might make life harder out here, it also makes me feel like *me*. Big, bold, brass, and unwilling to apologise for it. The streets might prefer the invisible, but I'll take being seen any day—even when it's more trouble than it's worth.

The Rules of the Bench

Every community has its rules, spoken or unspoken, and the bench was no different. To outsiders, it probably looked like chaos—a group of people lounging around, drinking, arguing, laughing, and existing in the kind of messy way that only the streets allow. But there were rules, and if you wanted to survive on the bench, you had to learn them fast.

The first rule of the bench was simple: respect the bench. It wasn't just a piece of wood and metal; it was home, sanctuary, and social hub all rolled into one. You didn't trash it, you didn't piss on it, and you sure as hell didn't try to claim it as your own. The bench belonged to everyone, and anyone who forgot that was quickly reminded. Usually with a few sharp words—or sharper fists, if the words didn't work.

The second rule: don't take what isn't yours. Theft was the fastest way to make enemies on the bench. It didn't matter if it was a lighter, a blanket, or a can of cider—if it wasn't yours, you left it alone. The problem, of course, was that the line between borrowing and stealing could get a little blurry when you were desperate.

I saw more fights break out over "borrowed" spice or "misplaced" bottles than I care to remember. The funny thing was, half the time, the missing item would turn up a day later, buried in someone's pocket or bag, forgotten in the chaos of survival. But by then, the damage was done, and the bad blood lingered.

The third rule: mind your business—until it's your business. The bench was a place of constant drama, and if you got involved in every argument, every fight, every lover's spat, you'd lose your mind. But there were times when you had to step in.

Like when Cat, the queen of the bench, was squaring off against a guy twice her size because he'd made the mistake of insulting her dog, Boss. Or when Benny, the old-timer with the harmonica, was cornered by a couple of drunk teenagers who thought it would be funny to steal his hat. Those were the moments when the bench came together, when the unspoken rules turned into unspoken alliances.

The fourth rule: share, but only if you can afford to. Sharing was complicated. On one hand, it was what kept us alive. A shared cigarette, a spare blanket, or a sip of someone's drink could mean the difference between a bad day and a manageable one.

But there were limits. You didn't share what you couldn't spare, and you didn't guilt people into giving. Everyone was struggling, and while generosity was appreciated, it wasn't owed. There was a kind of mutual understanding: give when you can, take only what you need, and don't push your luck.

The fifth rule: don't overstay your welcome. The bench was a shared space, and while most of us were regulars, there were always new faces drifting in and out. Some stayed for a day, others for weeks, but the unwritten rule was clear: if you weren't contributing to the vibe, you weren't welcome for long.

That didn't mean you had to be a saint—God knows noone was—but you had to pull your weight. That could mean sharing what little you had, telling a good story, or just not being an asshole. The bench had no patience for freeloaders or troublemakers.

And then there was the golden rule, the one that mattered most: take care of your own. The bench wasn't just a collection of individuals; it was a family, dysfunctional as it might have been. If someone was sick, we looked after them.

If someone was in trouble, we helped them out. It wasn't always pretty, and it wasn't always easy, but it was what kept us going. Out there, where the world felt like it was constantly trying to grind you down, the bench was a place where you could lean on others—even if just for a moment.

Of course, the rules weren't written in stone. They shifted, bent, and broke depending on the situation and the people involved. But they were there, guiding us in the chaos, reminding us that even in the most desperate circumstances, there's a way to live with some semblance of honor.

The bench had its enforcers, too. Cat was one of them, her right hook as effective as any lecture. Benny, with his harmonica and his sharp tongue, could shame someone into behaving better with a single well-placed insult. And then there was me, the odd one out, the guy who didn't quite fit the mold but somehow found his place anyway.

The rules of the bench weren't perfect. They didn't prevent every fight or fix every problem. But they gave us a framework, a way to navigate the madness and hold onto a shred of humanity. In a world that seemed determined to strip us of everything, the rules reminded us that we were still people, still capable of kindness, respect, and community.

And in the end, that's what the bench was about. It wasn't just a place to sit or a spot to pass the time. It was a microcosm of the world, messy and flawed but full of life. The rules weren't just rules—they were the glue that held us together, the thread that turned a group of strangers into something resembling a family. And for all its chaos, its fights, and its heartbreaks, the bench was home.

The Weight of the Bag

Out here, your life is reduced to what you can carry. Everything you are, everything you've been, everything you might need to survive—it all has to fit into one bag. Not a glamorous kit bag slung over the shoulder of a soldier on deployment, or a battered rucksack with a lifetime of adventures stitched into its fabric.

This is a bag stripped of romance and weighted with necessity. It's your lifeline and your anchor, the thing you'd die without and the thing that sometimes feels like it's killing you.

Choosing the bag is a strange ritual. It's not like picking out a new pair of trainers or splashing out on a fancy jacket. It has to be sturdy, able to survive the rain, the weight, the inevitable abuse of being dragged across the streets day in and day out. A cheap one won't last, and a flashy one invites the wrong kind of attention. You want something functional, inconspicuous, just enough to get by. A bag that says, *I'm here, but don't look too closely.*

What goes in the bag is even more important. Space is limited, so every item has to earn its place. Clothes come first, rolled tight to save room—always an extra pair of socks, maybe two if you're lucky. A toothbrush, toothpaste, and deodorant, if you can get your hands on them.

Then the essentials: a blanket, a waterproof jacket, maybe a spare tarp for emergencies. Food, if you have it, gets stuffed wherever it will fit, though it never lasts long enough to take up much space.

Some people carry little luxuries—a battered paperback, a photo from better days, a notebook filled with scrawled thoughts and half-finished ideas. These things don't serve any practical purpose, but they serve the soul. They're reminders of who you are, or who you used to be, or who you want to become again someday.

The bag also carries your survival tools. A lighter, even if you don't smoke, because fire can mean the difference between warmth and frostbite. A small first-aid kit, if you're lucky enough to have one. A bottle of water, always half-empty, always rationed like gold. Some people carry knives, but I never did. Weapons invite trouble, and trouble is the last thing you need.

Carrying the bag is its own challenge. It digs into your shoulders, rubs against your back, and grows heavier with every step. The straps fray, the zippers break, and the fabric wears thin. You patch it up when you can, but eventually, every bag gives out. When that happens, you find another, and the cycle begins again.

But the weight of the bag isn't just physical. It's the weight of knowing that this is all you have, that everything you own can be taken away in an instant. A careless moment, a thief in the night, or even a misplaced bag can strip you of everything. I've seen it happen—someone turning their back for a second, only to find their entire world gone when they turn around. Out here, losing your bag isn't just an inconvenience—it's a disaster.

The bag is also a symbol, a visible marker of your status. It tells the world who you are, or at least who they think you are. People glance at it and make their judgments. They see the worn straps, the bulging pockets, the stains from too many rainy nights, and they think they know your story. They don't. The bag isn't you, but it's the part of you the world sees first.

The bag isn't just a bag. It's a testament to survival, to resilience, to the ability to keep going even when the world feels like it's pressing down on you. It's a reminder that, no matter how heavy the weight, you're still here. Still carrying on. Still moving forward. And in a way, that makes it the most valuable thing you own.

"Brighton's allure lies in its ability to promise adventure, even in the mundane." —Travelogue Excerpt

Uncle Scott is Addicted to Adrenaline

If ever there was a man who lived like he was on a permanent time trial, it was Uncle Scott. Everything about him screamed momentum, like a machine that had been built to go fast and couldn't quite figure out how to stop. Even standing still, you got the sense he was always moving, always plotting, always chasing the next thrill, the next fix, the next story worth telling.

Scott was a living anachronism, a man who had once burned so brightly you could still see the afterimage, long after the light itself had faded. Back in the 90s, he was somebody—or so he told me, and I had no reason to doubt him. An international motocross rider, tearing up dirt tracks across Europe and sleeping with women who looked like they'd stepped off the cover of *Playboy*. Fast. Daring. Fearless.

I could picture it: Scott in his prime, lying over a dirt track, leathers clinging to a stocky frame, his grin a mix of cocky charm and reckless energy. The crowd roaring, his body one with the bike. He probably thought it would last forever. They all do, the men who live life in the fast lane, blind to the brick wall waiting at the end of it. He had the kind of charisma that could make you believe he'd done it all.

But life is cruel to men who burn too brightly. It doesn't let you stay in the sun for long. By the time I met him, his body was a crumbling monument to his glory days. His legs, once the source of his speed, were wrecked from injuries sustained on tracks and in scraps.

He could barely ride a pushbike now, let alone a motorbike. He walked with a peculiar skip in his step, almost jaunty, like a character dickensian chimney sweep. It made him seem lighter than he was, like he might float off if he didn't have the weight of his past dragging him down.

Scott's love affair with adrenaline hadn't faded, even as his body betrayed him. If he couldn't get his kicks on a bike anymore, he found other ways to chase the high. Shoplifting was his go-to thrill, and he treated it like an art form—or, at the very least, a sport.

Approaching it with the same reckless abandon he used to save for motocross. His weapon of choice was a bottle of white wine, always white, always the cheapest on the shelf. Before every "mission," he'd slap his thighs, adjust his collar, and mutter his battle cry in that perfect Cockney drawl: "Let's get the fucking job done."

Of course, he was terrible at it. He got caught constantly, arrested in more towns than I could count. "Every county's got a cell with my name on it," he once said, grinning like a schoolboy who'd just been caught nicking sweets. And yet, for all his failures, he never stopped. Shoplifting wasn't just about the wine; it was about the rush. It was about proving, if only to himself, that he could still outsmart the system—even if the system kept catching up to him.

Scott had a gift for picking fights he couldn't win. Maybe it was the adrenaline, or maybe it was the way he carried himself—like a man who used to own the room and hadn't realised the room had changed ownership. He'd square off with men half his age, blokes who could run circles around him if they wanted to. It was never about the outcome; it was about the act, the defiance, the moment of standing toe-to-toe and saying, "I'm still here."

I never once saw him win. Most of the time, he walked away limping, bloodied but unbroken. "Ain't about winning, mate," he told me once, a busted lip curling into a smirk. "It's about letting the bastards know you ain't afraid to lose."

For all his bravado, Scott had a way of stumbling into absurdity. There was the time he slept with a woman on the bench, a toothless vodka-and-spice addict who was old enough to be his mother.

When I asked him why, he shrugged with that mischievous glint in his eye and said, "I weren't even fucking looking." That was Scott in a nutshell—always landing himself in situations that sounded like punchlines to jokes only he could tell.

Uncle Scott wasn't a stranger to prison. "Done a few stretches," he'd say, as casually as someone might mention a gap year. He had the scars to prove it and the stories to make you laugh through the discomfort. His favorite tale was about making hooch in prison—fermenting bread in a hidden corner of his cell until it turned into something barely drinkable. "Tasted like piss, but it did the job," he said, his face lighting up at the memory. That phrase again. The job. To Scott, everything was a job—a mission to be completed, a challenge to be faced head-on.

Prison didn't scare him; nothing did, really. But it had left its mark. You could see it in the way he carried himself, in the way he never quite seemed at ease, even when he was laughing.

Scott was always full of life—messy, chaotic, sometimes maddening, but undeniably alive. That's why it terrified me the day I saw him collapse. He'd been drinking, of course—he always was—but this was different. He started convulsing, his body shaking so violently it looked like it might tear itself apart. Someone called an ambulance, and I just stood there, frozen, watching as the paramedics loaded him onto a stretcher.

Scott didn't say much, but when he did, it was pure gold. His Cockney accent turned every sentence into a one-liner, every story into a scene from a Guy Ritchie film. He had a way of making even the most mundane events sound like they belonged in a gangster flick.

After one particularly brutal arrest, he told me, "The coppers said, 'Scott, you're a fucking nuisance.' And I said, 'Yeah, but I'm your fucking nuisance.'"

It was impossible not to laugh, even as you wondered how much longer he could keep living this way.

For all his humor and swagger; his jokes and bravado; there was always something else lurking behind Scott's eyes. Pain, regret, maybe just the weight of a life lived too fast for too long. He never talked about it outright, but you could see it in the way he paused sometimes, staring off into the distance as if watching the ghost of a younger, faster version of himself.

But even with all that weight, Scott had a way of lifting others. He truly loved me, and he wasn't shy about saying it. "You've got a head on your shoulders, mate," he told me once, clapping me on the back with a hand that felt like it had been carved from oak. "Don't waste it."

Uncle Scott was a mess—a glorious, unforgettable mess. He was a relic of a different time, a man who refused to let go of the speed and chaos that had defined his youth. He wasn't a hero, not in the traditional sense, but he was a legend in his own right.

Every time I saw him, I couldn't help but smile. He made life on the bench feel a little less grim, a little more like something worth surviving. And for that, I'll always be grateful.

I don't know where Scott is now. Maybe he's in another town, chasing another thrill, telling his stories to someone new. Or maybe he's gone, his body finally giving out after years of pushing it too far. But wherever he is, I hope he's still moving, still chasing, still getting the fucking job done. I hope he's still fast, daring and fearless.

"Brighton's people have a resilience that matches the crashing waves on its shores, relentless and determined."—Julie Burchill

Ash

Ash. God, where do you even start with someone like him? He was like a grenade that had gone off years ago, and somehow the shrapnel was still tearing through the air, wreaking havoc long after it should've settled. Florida-born, Trinidad-hardened, Brighton-broken, he was the kind of guy who didn't just stumble into chaos—he carried it with him, a constant storm cloud that lit up everything around him with both electricity and destruction.

When you first met Ash, you might think he was some kind of eccentric tourist who'd lost his way. He had that wild, Beethoven hair, this careless, defiant grin, and an energy that felt like it should belong to someone on a beach in Miami, not freezing their arse off on a bench in Brighton. But then you got to know him—really know him—and the smile started to make sense. The chaos made sense. Everything made sense. And none of it did at the same time.

Florida wasn't just where Ash grew up; it was what made him. The sunburnt streets, the fried air thick with tension, the sound of boys laughing as they tossed basketballs against chain-link fences—those were his roots. His mum was English, his dad was a ghost, and Ash?

He was the white kid in the crowd of Black kids, sticking out like a sore thumb but fitting in all the same. If you know the streets, you know how rare that is. There's a saying that if you see a white guy hanging out with black guys, you should fear him because you have to wonder what he did to earn their respect.

That's Ash. It wasn't about race, not really. It was about respect. Ash had earned it somehow, some way. They gave him the N-word pass, the ultimate badge of belonging. And he wore it like armor, throwing it around in casual conversation like he'd been born to it.

But Florida wasn't a fairy tale, and the streets didn't let anyone off easy. Ash's friends were gangbangers, real ones, the kind you didn't cross unless you were suicidal or stupid. Ash was untouchable. His crew had his back, and he had theirs. They fought together, bled together, and shared a bond that was thicker than blood and far more dangerous. He was proud of it, too—proud in a way that only a kid desperate for belonging could be.

But Florida couldn't hold him. He got arrested one too many times, spent a couple of years in a Florida jail. A few years in the kind of prison you see in documentaries—the kind with heat that sucks the air out of your lungs and violence simmering just below the surface. He came out with two choices: stay and spiral deeper, or run. So he ran—to Trinidad, of all places.

Why Trinidad? "Because my boy was going," he told me once, as if that was reason enough. And for Ash, it probably was. He wasn't the type to overthink things.

At first, Trinidad was a fresh start. New place, new people, new chances. But the thing about a fresh start is that it doesn't matter where you go if you're carrying the same baggage. Ash didn't just fall back into old habits—he dove headfirst. Within months, he'd fallen in with a gang far more vicious than anything he'd known in Florida. These weren't kids with bruised knuckles and stolen bikes. These were grown men with machetes and no sense of mercy.

One night, he told me about the time they found out the owner of the bar they hung out in had raped his own daughter and gotten her pregnant. Ash's gang decided the man had to pay, but they didn't want Ash—the white boy—coming along. "You stay here," they told him, their voices as calm as if they were asking him to watch the bar while they went to the shop.

Ash was livid. He wanted to prove himself, to show he wasn't just some outsider tagging along. But when the gang came back, he understood. They'd beheaded the man, left his body as a message. "I didn't need to see that," he said, his voice quieter than I'd ever heard it. "I didn't need to be there."

That was Trinidad—chaos turned up to eleven. Ash tried to make it work, but the island chewed him up and spat him out.

Trinidad turned Ash into someone even he didn't recognise. He got a diamond grill while he was there, a piece of swagger that made him feel untouchable again, like the kid he used to be back in Florida. But swagger doesn't last long in a place like Trinidad. One day, a rival gang member stuck a pistol in his mouth and demanded the grill. Ash said no, because of course he did. That's who he was. The guy pressed the barrel harder against Ash's teeth, and that was the end of the argument.

"That grill was me," he told me later, shaking his head. "It wasn't about the diamonds. It was about—fuck, I don't know—it was mine."

When Trinidad became too much, Ash did what he always did—he ran. Back to Florida, back to his old crew, back to the chaos he knew. But Florida had changed. Or maybe Ash had.

Either way, it didn't take long for the cops to catch up with him again.

This time, they didn't just arrest him. They deported him. Ash had a British passport through his mum, but no American citizenship. So they put him on a plane to England—a country he barely remembered—and left him to fend for himself.

He landed in February, stepping off the plane with a suitcase full of shorts and a hoodie that wasn't built for the snow. "Thirty pairs of shorts," he told me, laughing. "Thirty fucking pairs. What was I thinking?"

Brighton wasn't kind to Ash. The council wouldn't house him—"No local connection," they said, as if that explained why a man with nothing couldn't even get a bed for the night. So he ended up on the bench, where I met him.

For all the shit he'd been through, Ash was one of the funniest people I'd ever met. He had this way of disarming you, making you laugh even when you didn't want to. He'd see a couple walking down the street and stop them dead in their tracks.

"Excuse me, is that your boyfriend?" he'd ask the woman, his face deadly serious. The boyfriend would tense up, bracing for a fight.

"Because he's fucking gorgeous."

Every time, without fail, they'd burst out laughing. That was Ash—a mix of charm, chaos, and mischief wrapped in a Florida drawl.

But there was always an edge to his humour, a sharpness that hinted at something darker. He'd crack jokes to cover the pain, to distract himself from the weight of everything he'd been through. You could see it in his eyes sometimes, that far-off look that said he was somewhere else entirely.

Ash looked like a young Beethoven—wild curls, sharp cheekbones, and a kind of manic energy that seemed to pour out of him like music. I used to give him clothes whenever I could, and he'd always be so grateful. "Thanks, man," he'd say, grinning as he pulled on a new hoodie or a clean pair of jeans. "You're the real MVP."

He was reckless, impulsive, and always getting himself into trouble. He picked fights he couldn't win, made jokes at the worst possible times, and seemed to attract chaos like a magnet. But he was also kind. And loyal. And so deeply human it hurt to watch sometimes. He'd often look me in the eye and say "Look TK, I'm a Dickhead but I'm not an asshole".

Ash wasn't built for the bench. He was too loud, too wild, too much. But he fit there in a way that no one else could. The bench was a place for people who didn't belong anywhere else, and Ash? Ash belonged everywhere and nowhere all at once. It was easy to underestimate Ash.

On the surface, he was all jokes and bravado, but under the his heavy backpack was a man carrying the weight of too many lives—gangs in two countries, jail time in one, and the wrong suitcase for this one. He wasn't supposed to be here, he'd tell me often, but that was more of an existential lament than a logistical one. England was a punishment, a frozen purgatory for someone who'd grown up in a world where shorts were a lifestyle, not a choice.

He was the hurricane that lit up the sky before tearing everything apart. And for all his flaws, for all his chaos, you couldn't help but love him. He was Ash. And that was enough.

"Brighton is the only place in England where a carnival atmosphere is almost perpetually maintained."—Henry James

A Siren Named Ella

Ella didn't belong on the bench. That much was obvious to everyone who saw her, and most of all, to her. She wasn't part of the scenery like the rest of us. She wasn't tethered to the place by addiction or poverty or circumstance. No, Ella came to the bench the way you might go to the theater—except here, the drama was raw and real, the kind of chaos that didn't wait for an intermission.

She was six feet of blonde hair, legs, and attitude, a walking anomaly that looked more at home on the cover of *Vogue* than surrounded by the rabble of the bench. Everything about her screamed *outsider*, from the way she dressed—long, tailored coats, knee-high boots, and designer scarves—to the way she carried herself, chin high, eyes cold. She was studying at Brighton University, second year, but whatever lectures or essays she had waiting for her always seemed less important than whatever chaos she could stir up here.

Ella had the kind of beauty that wasn't just distracting—it was weaponised. Men melted around her, turned into puddles of insecurity and lust, stumbling over their words like schoolboys. She thrived on it. You could see it in her eyes, that glint of satisfaction whenever some poor bastard made a fool of himself for her. She'd pull them in, wind them up, and then let them unravel.

It wasn't hard to figure out why. She'd been broken by men before. You could see it in the way she smiled—half cruel, half sad. It wasn't revenge exactly, but there was a kind of justice to it, a way of reclaiming power by taking it from others. She didn't just break men; she dismantled them. Piece by piece, word by word, until they didn't know who they were anymore.

There was this guy once, a boyfriend for a week, maybe less. He was a nice enough bloke, probably in over his head from the start. One day, I saw him punching himself in the face. Repeatedly. His fists slamming into his own skull as he muttered incoherently. A full-blown mental breakdown, right there on the street. When I tried to talk him down, he just kept repeating her name. Ella.

And she? She barely seemed to notice. She moved on as if he'd never existed, stepping over his wreckage like it was just another puddle on the pavement.

But for all her power, for all her beauty and cruelty, Ella was drowning too. You could see it if you looked closely enough.

The way her hands trembled when she thought no one was watching. The way her eyes would glaze over, unfocused and far away, as if she were somewhere else entirely.

She wasn't homeless, not technically, but she belonged to the bench in her own way. She came for the drama, sure, but there was something else, something deeper. The bench was where she could fall apart without anyone noticing. Or maybe it was where she could watch others fall apart and feel a little less broken herself.

Her addiction was obvious, though she wore it well, hidden behind layers of makeup and designer clothes. It was the kind of addiction that didn't scream or shout; it whispered, slowly pulling her under while she pretended not to hear.

One day, it nearly killed her.

It was a grey, unremarkable afternoon on the bench. The usual chaos was unfolding—arguments, laughter, people shouting over each other for no reason other than to be heard. And there, in the middle of it all, was Ella.

She was slumped against the back of the bench, her long legs sprawled out in front of her, her head tilted to the side. At first glance, she just looked tired, maybe bored. No one paid her any mind. The guys were too busy fighting over her, shouting insults and half-drunk declarations of love while she sat there silently, oblivious to it all.

But something wasn't right. Her chest barely moved. Her face was pale, almost grey.

I don't know what made me look closer—maybe it was the stillness, the kind of stillness that doesn't belong to someone like Ella. I shook her shoulder, gently at first, then harder when she didn't respond. Nothing.

"Shit," I muttered, standing up and shouting for someone to call an ambulance.

No one did. They were too busy yelling at each other, oblivious to the fact that the woman they were fighting over was slipping away right in front of them. So I called. My hands were shaking, my voice louder than it needed to be as I told the operator what was happening.

When the ambulance finally arrived, it was chaos. One of the guys fighting over her somehow ended up sitting in the front seat, grinning like an idiot as if he'd won a prize. The paramedics barely noticed; they were too focused on Ella, too busy saving her life to deal with the circus around them.

A few days later, I saw her again. She was back on the bench, dressed in her usual immaculate style, looking for all the world like nothing had happened.

She didn't even know I'd saved her life.

I didn't tell her. What would've been the point? Ella wasn't the type to say thank you, and I wasn't the type to expect it. She probably didn't even remember overdosing. Or maybe she did, and she just didn't care.

I sat down next to her, and for a moment, we didn't say anything. Just stared out at the street, watching the world go by. She lit a cigarette, her hands steady, her eyes cold and distant.

"You alright?" I asked eventually.

She glanced at me, a flicker of something—recognition, maybe—crossing her face. Then she smirked, exhaling a plume of smoke.

"Why wouldn't I be?"

And that was Ella. A hurricane wrapped in designer coats and nicotine, blowing through the bench like a force of nature. You couldn't save her, and you couldn't stop her. All you could do was watch.

"The sea is the same as it has been since before men ever went on it in boats, and the Brighton pier will be there tomorrow."—Ford Madox Ford

The Longest Bench

A man sits on a bench for long enough, and it changes him. It sharpens and blunts him at the same time, makes him harder in some ways, and softer in others. Afterwards, he moves on—because he has to—but he's not the same. Whatever else I do with my life—sit in a café with overpriced coffee, argue with a partner about something inconsequential, feel safe enough to forget what it means to survive—I'll always carry the bench with me.

It doesn't matter where I go or how far I walk. I carry it with me. The lessons I learned, the truths I uncovered, the things I saw that can never be unseen—they've rewired my soul. I can no longer look at the world without noticing the cracks, the ones people fall through so easily, the ones I once fell through myself.

The bench recalibrated me. It rewired my bullshit detector and ripped the rose-tinted glasses from my face. I see things differently now: the gap between the haves and the have-nots, the fragility of what we call "security," and the lies we tell ourselves about why some people end up in the gutter. It's not about bad choices. Not always. Often it's just a gust of bad luck that knocks you flat and leaves you staring at the sky while everyone else walks past without even glancing down.

Now, I can't unsee it. I'll never be able to unsee it. Every time I walk through a city, every time I catch the edge of a sleeping bag poking out from a doorway or see someone shivering under a bus shelter, I'll think of the bench. I'll think of Cat, Geordie, Ash, Ella, Uncle Stevie, and all the others. I'll think of myself.

The bench will go on, with or without me. New faces will come, old ones will leave, and the stories will keep unfolding. And maybe, just maybe, someone else will find what I found here—a strange kind of hope, forged in the unlikeliest of places.